Let's take a quick tour of the book.

Chapter 1 consists of definitions and basic results on polynomials. Some of the material, such as the discussion of graphing, is more advanced or abstract than the material in the rest of the book and need not be understood in detail on a first reading. A first reading gives orientation, and the chapter can be returned to as needed.

Chapter 2 treats the quadratic formula for the solution of degree two equations and examines it from several perspectives. The chapter closes with a selective history of the quadratic formula, focusing on the work of the ancient Babylonians nearly 4000 years ago and Al-Khwarizmi around 825 C.E.

The heart of the book is Chapters 3 through 6, in which we study polynomial equations of degrees three and four. The discovery of solutions to them by several mathematicians in sixteenth-century Italy represents a high point in mathematical history and was the most significant mathematical accomplishment in Europe for centuries. It also makes for a good story.

Chapter 3 introduces the solution to (reduced) cubic equations that has come to be known as Cardano's formula, in honor of Girolamo Cardano, a co-discoverer. Following the derivation of the formula and its use in some examples, we discuss graphs of cubic polynomials and the discriminant of a cubic, concluding with the dramatic story of the formula's discovery and publication.

One lesson of Chapter 3 is that Cardano's formula, when applied to certain cubic equations, expresses the solutions in terms of square roots of negative real numbers, obliging us to make sense of them. We do this in Chapter 4, learning how to work with an expansion of the system of real numbers known as the complex numbers and how to compute nth roots of complex numbers using trigonometry. The history of complex numbers is interesting. We close with a short account.

With complex numbers, we return in Chapter 5 to cubic equations, refining Cardano's formula and giving an algebraic description of the discriminant. We see how trigonometry can be used for the most difficult family of cubic equations to dispense with complex numbers altogether. Our historical review of cubics picks up where it left off two chapters earlier, as we learn how Cardano's successors came to terms with the appearance of complex numbers in his formula.

After our three-chapter-long struggle with the complexities of cubic equations, we enter smoother waters in Chapter 6, which is devoted to quartic equations. Out of respect for history, we begin with the original method of solving them, due to Cardano's assistant Lodovico Ferrari, but then turn to René Descartes' approach of a few decades later and Leonhard Euler's formula from a little over a century after that. All three methods require an intermediate step of solving a cubic equation that depends on the coefficients of the quartic. This is why the quartic waters are so much calmer; nearby, the cubic channel continues churning, but once we master the navigation of its shoals, the quartic provides little in the way of new obstacles. Euler's formula for the solution of a quartic equation leads to the calculation of the quartic's discriminant.

A recurring theme of the book is the abundance of information the coefficients of a polynomial encode about its roots. The quadratic formula is the first illustration, with more appearing throughout the discussion of cubic polynomials. The theme reaches its climax in Chapter 6, where we learn how the discriminant and other polynomial expressions in a quartic's coefficients can be used to determine the nature of its roots. The chapter concludes with a section inviting the reader to consolidate understanding of the material on cubic and quartic polynomials through a series of essay questions, followed by a historical section on the contributions of Descartes and Euler.

Chapter 7 is an introduction to some topics in the theory of polynomials of higher degree: quintic polynomials, the fundamental theorem of algebra, polynomial factorization, and symmetric polynomials. The level of difficulty rises, but full understanding is not the aim. The treatment is intended as a peek at what lies ahead, with results sketched, history surveyed, and little proved. We conclude on a highpoint that draws many of the chapter's elements together: a proof of the fundamental theorem of algebra that is essentially the one given by Pierre-Simon Laplace in 1795.

Acknowledgements: I thank my Math 497 students for working with early versions of this material. Their responses guided my selection of topics and my choice of emphasis. I am particularly grateful to Andrew Grzadzielewski for his continued encouragement.

Another student, Spencer Hubbard, read the manuscript and did all the exercises. His careful review and eye for detail led to many improvements, as well as enjoyable conversations. I thank Spencer in particular for his question about the roots of a quartic polynomial with zero discriminant, which motivated an expansion of Section 6.6.

The panel of reviewers that the Mathematical Association of America assembled to read an earlier version of the book did the most extraordinary of jobs, entirely transforming my vision of what it could be. No doubt I have fallen short, but the book is far better thanks to their advice and criticism, as well as later extensive comments by the copy editor, criticism of the history sections by an anonymous MAA reader, and transformation of my primitive diagrams by Beverly Ruedi. I am indebted as well to Don Albers and Jerry Bryce for their willingness to consider this project and for their ongoing support.

Exploring mathematics, even elementary mathematics, is a privilege, connecting us to fellow humans across millennia and cultures in our search for fundamental truth. (I hope this book illuminates these connections.) My greatest debt is to the members of my family, who have allowed and encouraged me to enjoy this privilege. My parents, to whom I have dedicated this book, arranged for me to arrive on a leap day, thereby inspiring my childhood interests in mathematics and astronomy. Their gift of Irving Adler's *The Giant Golden Book of Mathematics* [2] on my second birthday sealed my fate.

Contents

1

Polynomials

We will be studying polynomial equations throughout this book, especially those of degrees 2, 3, and 4. In this chapter, we introduce terminology and obtain some basic results that hold for all polynomials. The reader familiar with this material may wish to skip ahead. Others may wish on first encounter to read through this chapter's definitions and results, returning for a closer reading as they are used.

1.1 Definitions

What is a polynomial? We know that

$$3x^2 - 4x + 7$$

is one, as is

$$5x^{17} + 12x^{11} - 4x^7 + 13x^4 + x^3 - x + 113.$$

So are

$$\frac{4}{3}x^{100} + \frac{2}{5}x^{88} - 5$$

and

$$\sqrt{2}x^{333} - 2x^{200} + \frac{1}{3}x^{111} + x^4 - \pi^2.$$

But

$$x^4 + \sin x$$

is not a polynomial, and neither is

$$10^x.$$

(Why not? For now, we can agree that they certainly don't look like polynomials. We will return to this question in Exercises 1.12 and 1.14.)

The formal definition of a polynomial is: a *polynomial* is an expression of the form

$$a_n x^n + a_{n-1}x^{n-1} + a_{n-2}x^{n-2} + \cdots + a_2x^2 + a_1x + a_0,$$

where n is a non-negative integer and $a_n, a_{n-1}, \ldots, a_1, a_0$ are numbers. As indicated in the examples, the numbers might be integers, fractions, or more generally real numbers. (The real numbers are those numbers that mark positions on a line.) The numbers are called the *coefficients* of the polynomial. The coefficient a_0 is the *constant term* of the polynomial. Real numbers are themselves polynomials, the *constant polynomials*, with 0 called the *zero polynomial*.

For a non-zero polynomial, the largest integer d for which there is a non-zero coefficient a_d is called the *degree* of the polynomial. The expression $a_i x^i$ is the *degree i term* of the polynomial. By the definition of degree, the non-zero real numbers, viewed as polynomials, are degree 0 polynomials. By convention, 0 is said to have degree $-\infty$.

Ordinarily, when we write a polynomial as $a_n x^n + \cdots + a_1 x + a_0$, we anticipate that n is its degree and $a_n \neq 0$. Because this is not required by the definitions, we must exercise some care. Two polynomials $a_n x^n + \cdots + a_1 x + a_0$ and $b_m x^m + \cdots + b_1 x + b_0$ are *equal* if the corresponding coefficients are equal. More precisely, they have the same degree, d say, and $a_i = b_i$ for each $i = 0, 1, \ldots, d$. (If m or n is greater than d, then the indices a_i or b_j with $i > d$ or $j > d$ will be 0, by definition of degree.)

Exercise 1.1. What are the degrees of the four polynomials displayed at the start of the section?

Polynomials of low degree have special names: degree 1 polynomials are *linear* polynomials, degree 2 polynomials are *quadratic* polynomials or *quadratics*, degree 3 polynomials are *cubics*, degree 4 polynomials are *quartics*, and degree 5 polynomials are *quintics*.

A polynomial can be regarded as a function, taking on numerical values when numbers are substituted for x. This leads to the use of functional notation, with polynomials represented by expressions such as $f(x)$. Given a polynomial $f(x)$ and a number a, we write $f(a)$ for the number we get when we replace x by a. For example, suppose $f(x)$ is the polynomial

$$x^3 - 3x + 2.$$

Then

$$f(2) = 2^3 - 3 \cdot 2 + 2 = 8 - 6 + 2 = 4.$$

There is nothing special about x. We may write other letters for the variable (also called the *indeterminate*) of a polynomial, or any symbol in place of x. For example,

$$y^3 - 3y + 2$$

is a polynomial, as are

$$t^3 - 3t + 2$$

and

$$\clubsuit^3 - 3\clubsuit + 2.$$

Although written differently, they describe the same function.

Exercise 1.2. Is

$$\triangle^{37} + \sqrt[3]{2}\triangle^{11} - \pi\triangle + \frac{3}{7}$$

a polynomial? If so, what is its degree? If not, why not?

Given a polynomial $f(x)$ and a real number c, there is a natural way to form their product $c \cdot f(x)$: we multiply each coefficient of $f(x)$ by c. For example,

$$3 \cdot (4x^3 - 3x^2 + 2x + 8) = 12x^3 - 9x^2 + 6x + 24$$

and

$$\frac{1}{4} \cdot (4x^3 - 3x^2 + 2x + 8) = x^3 - \frac{3}{4}x^2 + \frac{1}{2}x + 2.$$

A non-zero polynomial is *monic* if the coefficient of its highest-degree term is 1. As we did in the example, we can multiply any non-zero polynomial by a suitable real number to obtain a monic polynomial: Given

$$a_n x^n + a_{n-1}x^{n-1} + \cdots + a_1 x + a_0$$

with $a_n \neq 0$, multiply by $1/a_n$.

Algebra has developed from the need to solve polynomial equations. A *polynomial equation* is an equation of the form

$$f(x) = 0,$$

with $f(x)$ a polynomial. A *solution* to this equation is a real number a satisfying

$$f(a) = 0.$$

The number a is also called a *root* of $f(x)$. For example,

$$x^5 - 3x^3 - 6x + 4 = 0$$

is a polynomial equation, and one can check that $x = 2$ is a solution.

Polynomial equations need not have solutions. For example,

$$x^2 + 1 = 0$$

doesn't. Substituting any real number for x, we obtain as its square a positive real number or 0. Adding 1 yields a positive real number, which therefore can't be 0.

In saying a polynomial equation has no solution, we are implicitly assuming that the domain of allowable solutions is the set of real numbers. In Chapter 4, we will introduce complex numbers, and we will find that equations such as $x^2 + 1 = 0$ have solutions in this expanded domain. It would be more accurate to say that $x^2 + 1 = 0$ has no real number solution rather than that it has no solution.

For any non-zero polynomial $f(x)$ and non-zero real number c, the polynomial equations

$$f(x) = 0$$

and

$$c \cdot f(x) = 0$$

have the same solutions. If a is a real number satisfying $f(a) = 0$, then $cf(a) = 0$ also, and similarly, if $cf(a) = 0$, then since c is non-zero, $f(a)$ must be 0. Therefore, whenever we want to solve a polynomial equation $f(x) = 0$ (with $f(x)$ non-zero), we can replace $f(x)$ by its associated monic polynomial and solve the resulting equation instead. We will make use of this elementary observation throughout the book.

1.2 Multiplication and Degree

Adding and multiplying polynomials is done in a first class in algebra. Nonetheless, it is important to state carefully what the operations are in general. The notation may seem complex, but the ideas and the results will be simple and should be familiar.

Two polynomials can be added and multiplied to obtain new polynomials. Addition is easy to define: we simply combine terms of like degree. Suppose $r(x)$ and $s(x)$ are polynomials given by

$$r(x) = a_n x^n + a_{n-1} x^{n-1} + \cdots + a_1 x + a_0$$

and

$$s(x) = b_n x^n + b_{n-1}x^{n-1} + \cdots + b_1 x + b_0.$$

(We do not assume that n is the degree of $r(x)$ or of $s(x)$, so a_n or b_n may be 0.) The *sum* $r(x) + s(x)$ is defined to be the polynomial

$$(a_n + b_n)x^n + (a_{n-1} + b_{n-1})x^{n-1} + \cdots + (a_1 + b_1)x + (a_0 + b_0).$$

The product is also as expected, but its description is more involved. Given real numbers a and b and non-negative integers m and n, we define the *product* of ax^m and bx^n to be

$$(ax^m) \cdot (bx^n) = abx^{m+n}.$$

If $m = n = 0$, we are multiplying the constant polynomials a and b, and their product is the usual product ab. This rule, combined with the distributive law, completely determines how two polynomials $r(x)$ and $s(x)$ are multiplied. If one polynomial is 0, the product is 0. Otherwise, their product $r(x) \cdot s(x)$ is the sum of the terms obtained by multiplying each term in $r(x)$ by a term in $s(x)$.

Exercise 1.3. Let us be more explicit about the coefficients of the product of two polynomials. Suppose $r(x)$ is a non-zero polynomial of degree m, with

$$r(x) = a_m x^m + a_{m-1}x^{m-1} + \cdots + a_1 x + a_0.$$

Suppose $s(x)$ is a non-zero polynomial of degree n, with

$$s(x) = b_n x^n + b_{n-1}x^{n-1} + \cdots + b_1 x + b_0.$$

(i) The coefficients a_m and b_n are non-zero. Why?

(ii) Show that the constant coefficient of $r(x)s(x)$ is $a_0 b_0$.

(iii) Show that the degree 1 coefficient of $r(x)s(x)$ is

$$a_0 b_1 + a_1 b_0.$$

(iv) Show that the degree 2 coefficient of $r(x)s(x)$ is

$$a_0 b_2 + a_1 b_1 + a_2 b_0.$$

(v) Show, for any non-negative integer k, that the coefficient of x^k in the product $r(x)s(x)$ is

$$a_k b_0 + a_{k-1} b_1 + a_{k-2} b_2 + \cdots + a_2 b_{k-2} + a_1 b_{k-1} + a_0 b_k.$$

(It is understood that $a_i = 0$ for $i > m$ and $b_j = 0$ for $j > n$.)

(vi) Check that the coefficient of x^{m+n} in $r(x)s(x)$ is $a_m b_n$.

(vii) Deduce Theorem 1.1.

Theorem 1.1. *Suppose m and n are non-negative integers, $r(x)$ is a polynomial of degree m, and $s(x)$ is a polynomial of degree n. Then the product $r(x)s(x)$ has degree $m + n$.*

Exercise 1.4. Practice multiplying polynomials. Determine the product of

$$x^{12} + 3x^7 + 4x^2 - 2$$

and

$$4x^5 - 3x^4 + 12x.$$

Write other polynomials and multiply them.

Given polynomials $f(x)$ and $r(x)$, the polynomial $r(x)$ *divides* the polynomial $f(x)$ if there is another polynomial $s(x)$ such that

$$f(x) = r(x)s(x).$$

We say that $r(x)$ is a *factor* of $f(x)$. As a consequence of Theorem 1.1, a factor of a non-zero polynomial $f(x)$ has degree less than or equal to the degree of $f(x)$.

Given two positive integers a and b, long division is an algorithm for dividing a into b, yielding a quotient q and a remainder r, with q and r non-negative and $r < a$. For example, if $a = 17$ and $b = 892$, long division produces the quotient $q = 52$ and the remainder $r = 8$. We learn this algorithm early in our mathematical studies. From the equation $b = aq + r$ and the fact that $r < a$, it follows that a divides b precisely when $r = 0$.

The long division process can be adapted to polynomials, as in high school algebra. Given two non-zero polynomials $a(x)$ and $b(x)$, we divide $a(x)$ into $b(x)$ to obtain a quotient $q(x)$ and a remainder $r(x)$, with $r(x)$ having lower degree than $a(x)$. The next exercise serves as review.

Exercise 1.5. Let $a(x)$ and $b(x)$ be non-zero polynomials. Let $q(x)$ be the quotient obtained by long division when one divides $a(x)$ into $b(x)$ and let $r(x)$ be the remainder.

(i) By the definition of quotient and remainder,

$$b(x) = a(x)q(x) + r(x).$$

(ii) Using the definition of divisibility, show that if $r(x) = 0$, then $a(x)$ divides $b(x)$.

(iii) Suppose instead that $a(x)$ divides $b(x)$. Show that $r(x)$ must be 0. (This requires a little care. Use the definition of divisibility to produce a polynomial $p(x)$ for which $b(x) = a(x)p(x)$. Compare with the equation in the first part of the exercise and use Theorem 1.1 to deduce that $r(x) = 0$.)

(iv) Conclude that to test if $a(x)$ divides $b(x)$, carry out long division and see if the remainder is 0.

(v) Decide if $x - 2$ divides $x^3 + 6x - 20$.

(vi) Decide if $x^2 + 2x + 1$ divides $x^5 + 4x^2 + 2$.

Let us collect some facts about multiplication and division that are analogues of familiar facts about number multiplication and division.

Theorem 1.2. *Suppose $f(x)$, $g(x)$, and $r(x)$ are polynomials.*

(i) If $f(x) \neq 0$ and $g(x) \neq 0$, then $f(x)g(x) \neq 0$.

(ii) If $f(x)g(x) = 0$, then at least one of $f(x)$ and $g(x)$ is 0.

(iii) If $r(x) \neq 0$ and $r(x)f(x) = r(x)g(x)$, then $f(x) = g(x)$.

Exercise 1.6. Prove Theorem 1.2.

(i) Prove the first part either directly from the definition of multiplication or from Theorem 1.1.

(ii) Explain why the second part follows from the first part.

(iii) Use the second part to prove the third part.

The third part of Theorem 1.2 says that when we have an equality of polynomial products, we can cancel a non-zero factor.

1.3 Factorization and Roots

The material in this section connects finding the roots of a polynomial $f(x)$ with finding degree 1 factors of $f(x)$. We will work in greater generality than we will need in our treatment of quadratic, cubic, and quartic equations. The generality allows us to see more clearly what the issues are, though we do more than is needed in later chapters.

Suppose we are given a polynomial $f(x)$ and a real number a such that $x-a$ divides $f(x)$. According to the definition of division, this means there is a polynomial $s(x)$ such that

$$f(x) = (x-a)s(x).$$

If we substitute a for x on both sides, we find that

$$f(a) = (a-a)s(a) = 0 \cdot s(a) = 0.$$

Let's record what we have just found as a theorem:

Theorem 1.3. *Given a polynomial $f(x)$ and a real number a, if $x-a$ divides $f(x)$, then $f(a) = 0$.*

Theorem 1.3 provides motivation for developing techniques to factor polynomials, as is done in a first algebra course. To factor the quadratic polynomial $x^2-13x+36$, recognize that $36 = 4 \times 9$ and $13 = 4+9$. This leads to the factorization

$$x^2 - 13x + 36 = (x-4)(x-9)$$

and shows that 4 and 9 are roots of $x^2 - 13x + 36$.

Factorization problems in a first algebra course usually deal with polynomials having integer coefficients. However, in this book we allow not just integers as coefficients but all real numbers. Thus, when presented with the polynomial x^2-2, we are interested in factoring it as a product

$$(x-r)(x-s)$$

for real numbers r and s. If we allow integers only, there is no such factorization, but with real numbers, we find the factorization

$$(x-\sqrt{2})(x+\sqrt{2}).$$

Theorem 1.3 then tells us (as is already evident) that $\sqrt{2}$ and $-\sqrt{2}$ are roots of x^2-2.

Given a polynomial $f(x)$ and a real number a, it is useful to know that if $x - a$ divides $f(x)$, then a is a root of $f(x)$. But will every root of $f(x)$ arise in this way? That is, if a is a root of $f(x)$, must $x - a$ divide $f(x)$?

Before answering this, let us review a point of logic. Many mathematical statements, such as Theorem 1.3, take the form, "If P is true, then Q is true," where P and Q are mathematical statements. For a statement of this type, the new statement "If Q is true, then P is true" is called its *converse*.

A statement and its converse are logically independent of each other: both may be true, one may be true while the other is false, or both may be false. Thus the fact that a statement is true sheds no light on the truth of its converse. For example, we know it is true that if an integer n is divisible by 10, then n is even. The converse states that if an integer n is even, then n is divisible by 10, and this is not true. Care is needed never to assume that the converse of a statement is true when the statement is true.

Theorem 1.3 is an example of a statement whose converse is true:

Theorem 1.4. *Given a polynomial $f(x)$ and a real number a, if $f(a) = 0$, then $x - a$ divides $f(x)$.*

Theorem 1.4 allows us to limit the possibilities for roots. If we can find all the factors of $f(x)$, we will be able to find its roots by examining the degree-one factors. The proof of Theorem 1.4 is given in the next exercise.

Exercise 1.7. Let $f(x)$ be a polynomial of degree n and let a be a real number. Let $x = y + a$.

(i) Suppose $f(x) = a_n x^n + \cdots + a_1 x + a_0$. Substitute $y + a$ for x to get a polynomial in y:

$$g(y) = a_n (y + a)^n + \cdots + a_1 (y + a) + a_0.$$

(ii) Use Theorem 1.1 to show for each positive integer i that $(y + a)^i$ has degree i.

(iii) Deduce that $g(y)$ has degree n and that therefore real numbers b_n, $b_{n-1}, \ldots, b_1, b_0$ exist for which

$$g(y) = b_n y^n + b_{n-1} y^{n-1} + \cdots + b_1 y + b_0.$$

(iv) From $y = x - a$, get a new expression for $f(x)$:

$$f(x) = b_n (x - a)^n + b_{n-1} (x - a)^{n-1} + \cdots + b_1 (x - a) + b_0.$$

(v) Let's determine b_0. Substitute a for x on both sides and show that

$$f(a) = b_0.$$

(vi) Conclude that

$$f(x) = f(a) + (x-a)\left(b_n(x-a)^{n-1} + \cdots + b_2(x-a) + b_1\right).$$

That is, given $f(x)$ and a, there is a polynomial $h(x)$ such that

$$f(x) = f(a) + (x-a)h(x).$$

(vii) Using this, prove Theorem 1.4.

Given a root a of a polynomial $f(x)$, we can ask how high a power of $x-a$ divides $f(x)$. The root a is called a *simple* root if $x-a$ divides $f(x)$ but no higher power does. It is a *multiple* or *repeated* root if a higher power of $x-a$ divides $f(x)$. The largest integer m such that $(x-a)^m$ divides $f(x)$ is called the *multiplicity* of a as a root of $f(x)$; that is, m is the multiplicity of a if $f(x)$ can be factored as $(x-a)^m g(x)$ for some polynomial $g(x)$ but not as $(x-a)^{m+1}h(x)$ for any polynomial $h(x)$.

1.4 Bounding the Number of Roots

It is natural to pursue the ideas of Section 1.3 further, and we shall do so. However, with two exceptions, we will not use the results of this section until Section 7.3. The exceptions are the proofs of Theorems 1.15 and 4.18, neither of which is essential for the development of the material in this book. Thus, this section can be omitted on a first reading.

We begin with an extension of Theorem 1.4:

Theorem 1.5. *Suppose $f(x)$ is a polynomial and $a_1, \ldots, a_k$ are k distinct real numbers that are roots of $f(x)$. Then the polynomial*

$$(x-a_1)(x-a_2)\ldots(x-a_k)$$

divides $f(x)$; that is, there is a polynomial $g(x)$ such that

$$f(x) = (x-a_1)(x-a_2)\cdots(x-a_k)g(x).$$

We will prove Theorem 1.5 in a sequence of exercises. First we comment on the use of the word *or* in logic or mathematics. Given two assertions P and Q, the statement "P or Q" means that at least one of the two holds.

The possibility that both hold is not excluded. For example, the statement "6 is even or 7 is even" is correct, since 6 is even, and the statement "6 is even or 8 is even" is correct also. If the statement "P or Q" is true but Q is not true, then P must be true.

We prepare for the proof of Theorem 1.5 with some polynomial divisibility facts.

Exercise 1.8. Let b and c be real numbers.

(i) Show that if $c \neq 0$, then the only polynomials that divide c are nonzero constant polynomials. (Use Theorem 1.1.)

(ii) Show that if $x - b$ divides $x - c$, then $b = c$. (If $x - c = (x - b)g(x)$ for some polynomial $g(x)$, what must $g(x)$ be?)

Exercise 1.9. Suppose $r(x)$ and $s(x)$ are polynomials and a is a real number. If $r(a)s(a) = 0$, then $r(a) = 0$ or $s(a) = 0$. (Why?) Use this and Theorem 1.4 to deduce that if $x - a$ divides $r(x)s(x)$, then $x - a$ divides at least one of $r(x)$ and $s(x)$.

We can now prove Theorem 1.5.

Exercise 1.10. Suppose $f(x)$ is a polynomial and $a_1, \ldots, a_k$ are k distinct real numbers that are roots of $f(x)$.

(i) Use Theorem 1.4 to show that there is a polynomial $g_1(x)$ satisfying

$$f(x) = (x - a_1)g_1(x).$$

(ii) Suppose $k > 1$. Since a_2 is a root of $f(x)$, use Theorem 1.4 and Exercise 1.9 to deduce that $x - a_2$ divides $x - a_1$ or $g_1(x)$.

(iii) By assumption, a_1 and a_2 are distinct. Use this and Exercise 1.8 to deduce that $x - a_2$ cannot divide $x - a_1$, and so therefore $x - a_2$ divides $g_1(x)$.

(iv) Conclude that there is a polynomial $g_2(x)$ satisfying $g_1(x) = (x - a_2)g_2(x)$, and therefore that

$$f(x) = (x - a_1)(x - a_2)g_2(x).$$

(v) Suppose $k > 2$. Make a similar argument to show that $x - a_3$ divides $(x - a_1)(x - a_2)$ or $g_2(x)$. Show that the first possibility can't hold, so that $x - a_3$ must divide $g_2(x)$. Conclude that there is a polynomial $g_3(x)$ satisfying $g_2(x) = (x - a_3)g_3(x)$, and therefore that

$$f(x) = (x - a_1)(x - a_2)(x - a_3)g_3(x).$$

(vi) Repeat $k - 3$ additional times to obtain a polynomial $g_k(x)$ satisfying

$$f(x) = (x - a_1)(x - a_2)\cdots(x - a_k)g_k(x).$$

An immediate consequence of Theorem 1.5 is the following important result about a polynomial and its roots:

Theorem 1.6. *Suppose $f(x)$ is a polynomial of positive degree n. Then $f(x)$ has at most n distinct roots.*

Exercise 1.11. Prove Theorem 1.6 using Theorem 1.5 and a comparison of degrees.

We noted in Section 1.1 that $\sin x$ is not a polynomial. This now follows as a consequence of Theorem 1.6.

Exercise 1.12. Deduce from Theorem 1.6 that $\sin x$ cannot be equal to a polynomial. (Hint: How many solutions are there to $\sin x = 0$?) Show similarly that $\cos x$ and $\tan x$ are not polynomials.

1.5 Real Numbers and the Intermediate Value Theorem

In this section, we temporarily leave algebra to introduce the intermediate value theorem. We will make no effort to prove it, since the proof depends on foundational results about the real numbers whose development would take us too far afield. But the theorem is easily stated, matches our intuition, and will be useful to have for reference, as is explained in more detail at the start of Section 1.6.

The primary question we wish to answer about a polynomial $f(x)$ is whether it has a root. Viewing $f(x)$ as a function, we are asking whether $f(x)$ takes on the value 0 when a real number is substituted for x, or equivalently, whether the graph of $y = f(x)$ crosses the x-axis.

The answer to this question depends on an understanding of the properties of the real numbers. To see why, let's consider the simplest nontrivial example for which we may find ourselves searching for a root. Every degree-one polynomial $ax + b$ has a root (what is it?), so we may turn to degree 2. Given a positive real number r, does the polynomial $x^2 - r$ have a root? Equivalently, does r have a square root? Let's be specific: does 2 have a square root?

The first numbers we learn about are the *integers*: $0, 1, 2, \ldots$, and their negatives. Next we learn about *rational numbers*, the ratios a/b of two integers a and b, with $b \neq 0$. No integer has a square equal to 2. Nor is there a rational number whose square is 2. This is a famous discovery often credited to Pythagoras some 2500 years ago. You can rediscover it in the following exercise.

Exercise 1.13. Prove that 2 does not have a rational square root.

(i) Suppose a and b are positive integers such that a/b is a square root of 2. Assume further that a/b is a reduced fraction; that is, no positive integer divides a and b other than 1. Square and clear denominators to obtain $a^2 = 2b^2$.

(ii) Show that 2 must divide a.

(iii) Show that 2 must divide b.

(iv) This is a contradiction, proving that a and b can't exist.

We learn early in our mathematical education that 2 does have a square root, and that it is given by a decimal expansion that we can describe to as many terms as we wish. Squaring 1 and 2, we find that 1^2 is less than 2 and 2^2 is greater than 2, so we anticipate that the square root, if it exists, lies between them. With further testing, we find that 1.4^2 is too small and 1.5^2 is too large, so the square root should lie between them. Iterating, we get better and better approximations to $\sqrt{2}$. For instance, after a few minutes work (or almost no work at all, when armed with a calculator), we get the approximation 1.4142, whose square is 1.99996164, or the poorer approximation 1.4143, whose square is 2.0002449, and we conclude that the square root lies between the two numbers. Continuing, we obtain as many terms as we wish of a decimal expansion that we take to be the desired square root.

The process of determining a square root of 2 in this way may seem natural, but its rigorous justification requires laying a proper foundation for the definition and properties of the real numbers. This book is not the place to lay such a foundation. We will instead review some essential facts about real numbers and functions on real numbers, along with their consequences for the study of polynomials. Many introductory real analysis texts cover this material. Our standard reference will be Steven G. Krantz's *Real Analysis and Foundations* [38].

We must start with a formal definition of the real numbers. We first learn that a real number consists of an integer or whole number part and a fractional part given by a possibly infinite decimal expansion. By moving from rational numbers to this larger class, we fill the holes. Any real number can be approximated by a rational number as closely as desired: we truncate the real number's decimal expansion farther and farther out to get better and better approximations. But only by allowing the full infinite expansion do we succeed in filling the holes, assigning a real number to every point on the number line. We will omit a detailed development of this process, relying instead on the reader's intuition about real numbers as given by decimal expansions. (Krantz provides a treatment in two parts, introducing the properties of the set of real numbers and stating a theorem that such a set of numbers exists [38, pp. 58–62], then providing the technical proof of the theorem in an appendix [38, pp. 71–73].)

Given two real numbers $a < b$, we introduce two sets of real numbers associated to a and b, the *closed interval* $[a, b]$, which consists of all numbers x satisfying $a \leq x \leq b$, and the *open interval* (a, b), consisting of all numbers x satisfying $a < x < b$.

Once the real numbers are introduced, we can study properties that sets of real numbers may satisfy, such as *connectedness*. We will omit the formal definition, but from it we can prove two basic results [38, pp. 145–147]: open and closed intervals are connected, and any connected set of real numbers that includes the numbers $r < s$ will contain the entire closed interval $[r, s]$.

We also introduce the notion of *continuity*, for functions defined on the real numbers or a suitable subset. We will not go into the details (see [38, pp. 159–164]). Informally, a continuous function is one whose graph doesn't skip around or have gaps. From the formal definition, we can verify that any constant function $f(x) = c$ is continuous, as is $f(x) = x$. We can also prove that sums and products of continuous functions are continuous, from which it follows that any polynomial function is continuous. This is sufficiently important to record as a theorem:

Theorem 1.7. *A polynomial with real coefficients, regarded as a function on the real numbers, is continuous.*

Another key result in the study of continuous functions is the theorem [38, pp. 169–170] that a continuous function takes a connected set of real numbers to a connected set. Combining this with the result that a connected set of real numbers containing r and s contains all the numbers in between, we can easily deduce the intermediate value theorem [38, p. 170].

Theorem 1.8 (intermediate value theorem). *Let $a < b$ be real numbers and let f be a continuous function defined on the closed interval $[a, b]$. Given a real number d between $f(a)$ and $f(b)$, there is a real number c in $[a, b]$ such that $f(c) = d$.*

Geometrically, the intermediate value theorem tells us that if the graph of a continuous function goes through two different heights, then the graph goes through every height in between as it passes from one height to the other. Although the theorem is fundamental to the development of calculus, its proof is usually deferred to an advanced calculus or real analysis course.

Since polynomials are continuous functions, the intermediate value theorem applies to them, yielding the following result.

Theorem 1.9 (intermediate value theorem for polynomials). *Let $a < b$ be real numbers and let $f(x)$ be a polynomial. Given a real number d between $f(a)$ and $f(b)$, there is a real number c in $[a, b]$ such that $f(c) = d$.*

As an example of the intermediate value theorem in action, given a positive real number d that is not the square of an integer, we can find non-negative integers $a < b$ with $a^2 < d < b^2$. By the intermediate value theorem, there is a real number c satisfying $a < c < b$ and $c^2 = d$. So it is a consequence of the intermediate value theorem that every positive real number has a positive square root. This is sufficiently important to record as a theorem.

Theorem 1.10. *Given a positive real number d, there exists a real number c such that $c^2 = d$, so every positive real number has a square root.*

The same argument works when we replace the exponent 2 by any positive integer n, showing that every positive real number has a positive nth root.

Here is another application of the intermediate value theorem, stated for polynomials although it holds for any continuous function:

Theorem 1.11. *Let $a < b$ be real numbers and let $f(x)$ be a polynomial for which $f(a) < 0$ and $f(b) > 0$. Then there is a real number c between a and b such that $f(c) = 0$.*

In terms of graphs, the theorem tells us that if there is a point on the graph of $y = f(x)$ that is below the x-axis and another point that is above the x-axis, then there must be a point of the graph on the x-axis.

1.6 Graphs

In Section 1.5, we introduced some results on real numbers. Here, we will continue our excursion into realms beyond algebra, this time turning to analytic geometry to study the shape of the graph of a polynomial function. We will give merely an overview. Calculus gives the results of this section as easy exercises. Some of the results can be obtained by more elementary considerations, albeit at the cost of a little more work, but either way, foundational theorems on the construction and properties of the real numbers are essential for the proofs.

We will use the results of this section in Sections 2.4, 3.3, and 3.4, where the discriminants of quadratic and cubic polynomials are studied. In Sections 4.2 and 5.3, the results will be proved again algebraically, so that their derivation does not depend on the results of this section. Nonetheless, the approach to discriminants using graphs in Sections 2.4 and 3.4 provides additional insight and intuition.

The polynomial functions whose graphs are the simplest are the powers of x. We know that the graph of $y = x$ is the line forming a 45° angle with the x-axis. Suppose n is an integer greater than 1. For positive real numbers $a < b$ and $r < s$, the inequality $ar < bs$ holds. Setting $r = a$ and $s = b$, we find that $a^2 < b^2$. For a positive integer n, we can repeat the argument n times to obtain $a^n < b^n$. This tells us that the graph of $y = x^n$ is always increasing in height as x increases, for $x \geq 0$.

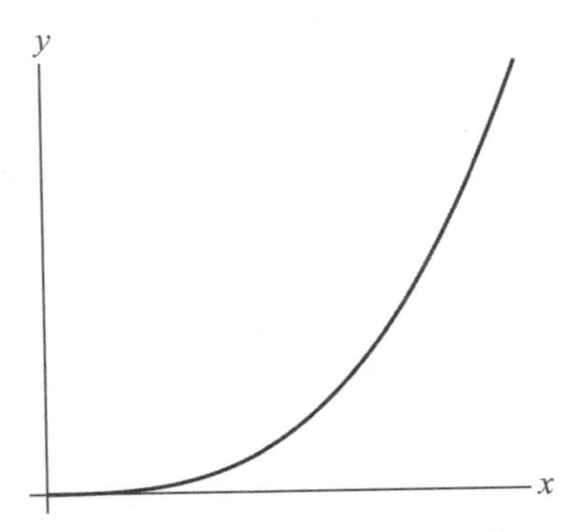

Figure 1.1. Graph of $y = x^n$

We also know that the graph of $y = x^n$ increases in height without bound; that is, for any positive number N, there is a real number r for which $r^n > N$. (For $N > 1$, the inequality will hold if we take r to be an integer greater than N.) We can use the intermediate value theorem, Theorem 1.9, to deduce that as x increases, with $x > 0$, the function x^n assumes all possible positive real number values. This leads to the next theorem, and to the sketch in Figure 1.1 illustrating the behavior of the graph of $y = x^n$ to the right of the y-axis.

Theorem 1.12. *Let n be an integer greater than* 1.

(i) For $x \geq 0$, the graph of $y = x^n$ is strictly increasing; that is, $a^n < b^n$ for any real numbers a and b with $0 \leq a < b$.

(ii) The graph of $y = x^n$ goes through all non-negative heights; that is, for any non-negative real number s, there is a non-negative real number r satisfying $r^n = s$.

One feature of the graph in Figure 1.1 that requires calculus to justify is the way the shape has been drawn. It's what is called *concave up*. Those familiar with calculus will recognize that this follows from the fact that for $x > 0$, the second derivative $n(n-1)x^{n-2}$ of x^n is positive.

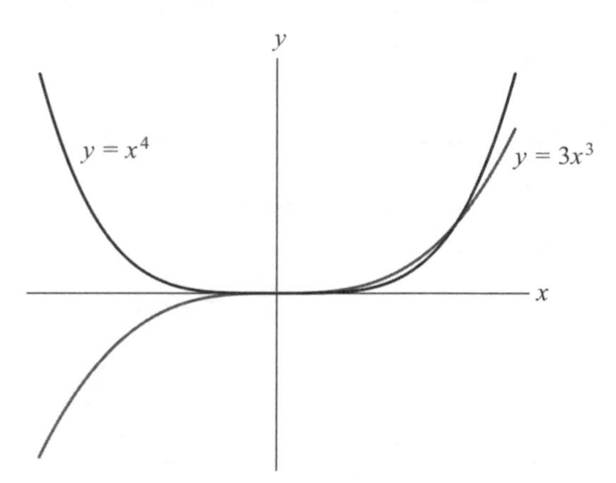

Figure 1.2. Graphs of even and odd powers of x

We have discussed x^n only for $x \geq 0$. If $x < 0$, then $(-x)^n = x^n$ for n even. This tells us that the graph to the left of the y-axis is a mirror image of the graph to the right. Thus, as x increases, the graph falls steadily from arbitrarily large heights to $(0, 0)$, then rises arbitrarily high. If instead n is odd, then $(-x)^n = -x^n$, so that the graph to the left of the y-axis is obtained from the graph to the right by taking the mirror image across the y-axis and then reflecting again across the x-axis. We see from this that the graph rises steadily as x increases, from arbitrarily low heights to $(0, 0)$ and then onwards to arbitrarily high heights. This is illustrated in Figure 1.2.

The graph of a general polynomial will be more complicated, but some features remain unchanged from that of a simple power of x. For example, consider the graph of the polynomial $x^7 - 6x^5 + 11x^3 - 6x$ depicted in Figure 1.3. Like the graph of an odd power of x, it rises to infinity to the right and drops to infinity to the left. The complicating feature is that it rises and falls in between. Counting, we find that it makes six turns on its way from minus infinity to infinity. (Those familiar with calculus will recognize that a seventh-degree polynomial can have no more turns, though it may have fewer. For instance, x^7 has no turns.)

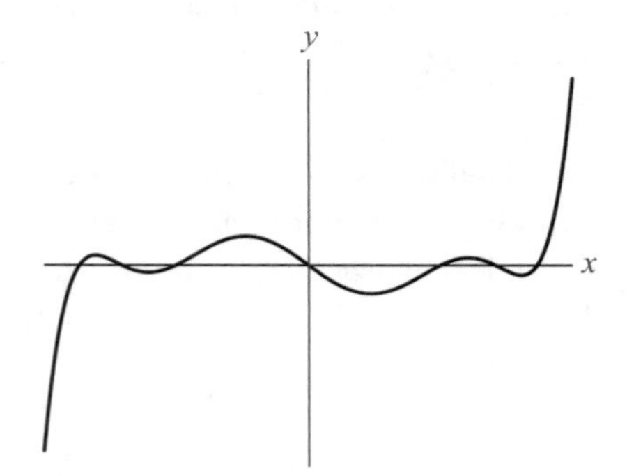

Figure 1.3. Graph of $y = x^7 - 6x^5 + 11x^3 - 6x$

Let's discuss the general picture. Suppose that $f(x)$ is a monic polynomial of positive degree n,

$$x^n + a_{n-1}x^{n-1} + \cdots + a_1 x + a_0.$$

We can factor x^n out to obtain

$$f(x) = x^n \left(1 + \frac{a_{n-1}}{x} + \cdots + \frac{a_1}{x^{n-1}} + \frac{a_0}{x^n}\right).$$

For x very large, the terms other than 1 inside the parentheses will be small in absolute value and $f(x)$ will be close to x^n. Using estimate arguments typical of calculus, we can convert this idea into a proof that there is a positive real number N such that for $x \geq N$, the graph of $f(x)$ rises steadily, and without bound. Since the graph rises without bound, we can further conclude by the intermediate value theorem that the graph goes through every height above $f(N)$. For x negative but large in absolute value, the same argument allows us to show, if n is even, that the graph of $f(x)$ decreases from unbounded heights as x increases through negative values and, if n is odd, that the graph increases from arbitrarily low levels. We summarize what we have found in the next theorem.

Theorem 1.13. *Let $f(x)$ be a monic polynomial of positive degree n. There exists a positive real number N for which $f(x)$ satisfies:*

(i) Given real numbers $b > a \geq N$, we have $f(a) < f(b)$.

(ii) For any real number $s > f(N)$, there is a real number $r > N$ such that $f(r) = s$. That is, the graph of $f(x)$ increases for $x \geq N$ and goes through all heights above $f(N)$.

(iii) Given real numbers $a < b \leq -N$, we have $f(a) > f(b)$ if n is even and $f(a) < f(b)$ if n is odd.

(iv) If n is even, then for any real number $s > f(-N)$, there is a real number $r < -N$ such that $f(r) = s$; if n is odd, then for any real number $s < f(-N)$, there is a real number $r < -N$ such that $f(r) = s$. That is, if n is even (odd), the graph of $f(x)$ decreases (increases) for $x \leq -N$ and goes through all heights above (below) $f(-N)$.

When n is even, since the graph of $f(x)$ falls, then rises, it may never cross the x-axis. This is the graphical way of stating that an even-degree polynomial may have no roots. For instance, as we have seen, $x^2 + 1$ has no roots.

In contrast, by Theorem 1.13, an odd-degree polynomial will assume negative values for x negative of large absolute value and positive values for x positive and large. By Theorem 1.11, the polynomial must assume the value 0 somewhere. This is important enough to record as a theorem:

Theorem 1.14. *Let $f(x)$ be a polynomial of odd degree. Then there is a real number r such that $f(r) = 0$. That is, every odd-degree polynomial has a root.*

Theorem 1.13 tells us that the graph of a degree-n polynomial $f(x)$ behaves like the graph of x^n in its extremes. Away from them, the graph may behave differently from the graph of x^n. It may shift from rising to falling to rising multiple times, as illustrated by the seventh-degree polynomial graph of Figure 1.3.

A graph changes from falling to rising or rising to falling at what are called turning points. Let us define this precisely. A *local maximum* of a function $f(x)$ is a point $(a, f(a))$ on its graph with the property that there exists a positive number r such that every c in the open interval $(a-r, a+r)$ satisfies $f(c) \leq f(a)$. Similarly, $(b, f(b))$ is a *local minimum* if there is a positive number s such that every c in the open interval $(b-s, b+s)$ satisfies $f(b) \leq f(c)$. A *turning point* for $f(x)$ is a point that is either a local maximum or a local minimum.

An example of a turning point is the point $(0, 1)$ on the graph of $f(x) = x^2 + 1$, for which it is a local minimum. In fact, since $f(x) > 1$ for any $x \neq 0$, the point $(0, 1)$ is more than a local minimum. It's what is called a *global minimum*.

Readers familiar with calculus will know, given a differentiable function $y = f(x)$, that for $(a, f(a))$ to be a turning point, $f'(a)$ must equal 0. Thus, the turning points, if there are any, will be among the points where the derivative vanishes. The converse does not hold: $f'(a)$ may equal 0 without $(a, f(a))$ being a turning point. An example is x^3, whose derivative at 0 is 0, but $(0, 0)$ is not a turning point, since the function $f(x) = x^3$ is always increasing. (The graph is rising when $x \neq 0$ but flat at $x = 0$.)

If $f(x)$ is a polynomial of positive degree n, we learn in calculus how to compute its derivative $f'(x)$, and find that $f'(x)$ is itself a polynomial, of degree $n-1$. By Theorem 1.6, $f'(x)$ can have at most $n-1$ distinct roots; that is, the derivative of $f(x)$ vanishes at at most $n-1$ points. Since these are the only candidates for turning points, we obtain the next theorem.

Theorem 1.15. *A polynomial of positive degree n has at most $n-1$ turning points.*

We can use Theorems 1.13 and 1.15 to see again that the sine function cannot be a polynomial, and to see that the exponential function 10^x isn't a polynomial.

Exercise 1.14. This exercise assumes familiarity with $\sin x$ and 10^x as functions defined for all real values of x.

(i) The graph of $y = \sin x$ has the shape of an infinite wave, oscillating between peaks of height 1 and valleys of height -1. Describe the turning points of $\sin x$ and deduce that it has infinitely many local minima and infinitely many local maxima.

(ii) Deduce from Theorem 1.15 that $\sin x$ cannot be a polynomial.

(iii) Use Theorem 1.13 to give an alternative proof that $\sin x$ cannot be a polynomial.

(iv) For $x < 0$, we have $0 < 10^x < 1$. Deduce from Theorem 1.13 that the function $y = 10^x$ cannot be a polynomial.

2

Quadratic Polynomials

The heart of this book is the study of solutions to cubic and quartic equations, which we will begin in Chapter 3. This chapter is devoted to quadratic equations. Even though they are familiar from a first algebra course, a close look is warranted, as a warmup before we tackle the greater difficulties of cubic and quartic equations and to introduce themes that will recur as we study cubics and quartics.

The general quadratic equation has the form

$$ax^2 + bx + c = 0,$$

for real numbers a, b, and c, with $a \neq 0$. The *quadratic formula* for the solutions of this equation takes the form

$$x = -\frac{b}{2a} \pm \frac{\sqrt{b^2 - 4ac}}{2a}.$$

Since $a \neq 0$, we are free to divide both sides of the equation $ax^2 + bx + c = 0$ by a before searching for solutions. This amounts to setting $a = 1$ and studying quadratic equations of the form

$$x^2 + bx + c = 0.$$

We will take this approach throughout this chapter. The quadratic formula then takes the simpler form

$$x = -\frac{b}{2} \pm \frac{\sqrt{b^2 - 4c}}{2}.$$

We will obtain the quadratic formula (in this simpler form) by three different approaches, and turn to some history at the end.

2.1 Sums and Products

Our initial approach to the quadratic formula depends on solving a seemingly different problem, the determination of two numbers given their sum and product. Let's begin with a related but simpler problem.

Given the sum and difference of two numbers, can we determine the numbers? For instance, suppose I tell you that I am 36 years older than my son and that our ages sum to 84. Can you determine how old we are?

Exercise 2.1. Solve the age problem.

Exercise 2.2. Solve the general form of the age problem. Suppose r and s are unknown numbers, with $u = r + s$ and $v = r - s$. Determine r and s in terms of u and v.

We see from the solution to Exercise 2.2 that we can determine two numbers from their sum and difference. It is also possible to determine two numbers from their sum and product. Doing so requires the calculation of a square root, which leads to an ambiguity in sign. But it turns out to be harmless, since the two solutions it yields are the two we seek.

Exercise 2.3. Given two numbers r and s, suppose that $u = r + s$ and $v = r - s$, as in Exercise 2.2. Set $p = rs$.

(i) Verify that

$$(r + s)^2 - (r - s)^2 = 4rs,$$

so

$$u^2 - v^2 = 4p.$$

(ii) Write this as

$$v^2 = u^2 - 4p$$

and take square roots to obtain a formula for v in terms of u and p, correct up to an ambiguity in the sign.

(iii) Using Exercise 2.2, find r and s in terms of u and p and conclude that Theorem 2.1 holds.

Theorem 2.1. *Given two real numbers r and s, let u be their sum and p their product. Then r and s are the two quantities*

$$\frac{u}{2} \pm \frac{\sqrt{u^2 - 4p}}{2}.$$

Because of the uncertainty in sign arising from taking a square root, we can't specify which expression in Theorem 2.1 is r and which is s. However, this doesn't matter, since we are interested only in identifying the pair of numbers r and s, not in knowing which is which.

Theorem 2.1 is essentially the quadratic formula. Before we see this, we will obtain a general result about quadratic polynomials that is a consequence of Theorem 1.4, which stated that if a polynomial $f(x)$ (of any degree) has a root a, then $x - a$ divides $f(x)$.

Exercise 2.4. Let b and c be real numbers and let $f(x)$ be the polynomial $x^2 + bx + c$.

(i) Suppose $f(x)$ factors as $(x-r)(x-s)$ for distinct real numbers r and s. Show that r and s are the only roots of $f(x)$. (Hint: Suppose t is a root. Substitute it for x.)

(ii) Conversely, show that if distinct real numbers r and s are roots of $f(x)$, then $f(x) = (x-r)(x-s)$. (Hint: Use Theorem 1.5.)

(iii) Suppose $f(x)$ factors as $(x-r)^2$. Show that r is the only root of $f(x)$.

(iv) Conclude by proving Theorem 2.2.

Theorem 2.2. *Let $f(x) = x^2 + bx + c$ for real numbers b and c. Exactly one of the three possibilities occurs:*

(i) $x^2 + bx + c$ has two distinct real roots r and s, and factors as

$$(x-r)(x-s).$$

(ii) $x^2 + bx + c$ has only one real root r, and factors as $(x-r)^2$.

(iii) $x^2 + bx + c$ has no real roots, and cannot be factored as a product of degree 1 polynomials.

Given a quadratic polynomial $x^2 + bx + c$ with real coefficients b and c and with real roots r and s, to find a formula for r and s in terms of b and c, it would suffice by Theorem 2.1 to find formulas for $r + s$ and rs in terms of b and c. We could then recover r and s from the values for $r + s$ and rs. This turns out to be easy to do.

Exercise 2.5. Suppose that the polynomial $x^2 + bx + c$ has real roots r and s, distinct or equal. Theorem 2.2 yields a factorization of $x^2 + bx + c$ as

$$(x-r)(x-s).$$

(i) Multiply the expression $(x - r)(x - s)$ to show that

$$r + s = -b$$

and

$$rs = c.$$

(ii) Use Theorem 2.1 to obtain expressions for r and s in terms of b and c. This proves Theorem 2.3.

Theorem 2.3 (Quadratic Formula). *Let b and c be real numbers and suppose $x^2 + bx + c$ has real roots r and s (distinct or coincident). Then r and s are the two quantities*

$$-\frac{b}{2} \pm \frac{\sqrt{b^2 - 4c}}{2}.$$

We see from Theorem 2.1 and Exercise 2.5 that the quadratic formula is the expression for the roots of $x^2 + bx + c$ in terms of their sum $-b$ and their product c.

Let us consider how the quadratic formula of Theorem 2.3 provides a solution for quadratic equations. In the simplest case, with $b = 0$, the equation has the form

$$x^2 + c = 0$$

and the quadratic formula tells us that the solution is $x = \pm\sqrt{-c}$. If $c > 0$, then $-c$ has no square roots and there is no solution. If $c < 0$, then $-c$ is positive and has two square roots. The quadratic formula does not tell us how to compute them. It tells what we already know, that the solutions to the equation are the square roots. Thus, rather than regarding the formula as a way to solve an arbitrary quadratic equation, we should view it as a reduction technique. It gets us to the point of having to do a square root calculation, then leaves us on our own. As we saw in Section 1.5, calculating square roots is not a problem of algebra.

2.2 Completing the Square

We now turn to a second approach to deriving the quadratic formula, completing the square.

Exercise 2.6. Solve the quadratic equation

$$x^2 + 10x - 39 = 0.$$

(i) Write the equation as

$$x^2 + 10x = 39.$$

(ii) Divide 10 by 2 to get 5 and add 5^2 to both sides to obtain

$$x^2 + 10x + 5^2 = 39 + 5^2.$$

The left side is the square of $x + 5$.

(iii) Write the equation as

$$(x + 5)^2 = 64,$$

take square roots on both sides, and conclude that the equation has solutions $x = 3$ and $x = -13$.

We can now take up the general case.

Exercise 2.7. Let a, b, and c be real numbers.

(i) Verify that

$$(x + a)^2 = x^2 + 2ax + a^2.$$

(ii) Deduce that the polynomial

$$x^2 + bx + \frac{b^2}{4}$$

is the square of a degree one polynomial.

(iii) Rewrite

$$x^2 + bx + c$$

by adding and subtracting $b^2/4$ and find that solving the equation $x^2 + bx + c = 0$ is equivalent to solving

$$(x + b/2)^2 = d/4$$

for a suitable real number d. Write d in terms of b and c.

(iv) Deduce for $d < 0$ that there is no solution to $x^2 + bx + c = 0$.

(v) Deduce for $d = 0$ that $x^2 + bx + c$ factors as $(x + b/2)^2$ and the one solution to $x^2 + bx + c = 0$ is $x = -b/2$.

(vi) Deduce for $d > 0$ that there are two solutions to $x^2 + bx + c = 0$, involving the square root of d. Write the solutions in terms of b and c, giving Theorem 2.4.

Theorem 2.4. *Let b and c be real numbers.*

(i) If $b^2 - 4c < 0$, then the equation

$$x^2 + bx + c = 0$$

has no solutions.

(ii) If $b^2 - 4c = 0$, then the only solution of $x^2 + bx + c = 0$ is $x = -b/2$.

(iii) If $b^2 - 4c > 0$, then the equation $x^2 + bx + c = 0$ has two solutions, given by

$$x = -\frac{b}{2} \pm \frac{\sqrt{b^2 - 4c}}{2}.$$

The quadratic formula takes on a simpler form if we alter how we write the coefficients in our initial quadratic equation. Given the real numbers b and c, let $B = b/2$ and $C = -c$. Then $b = 2B$, $c = -C$, and

$$x^2 + bx + c = 0$$

can be written as

$$x^2 + 2Bx - C = 0,$$

or

$$x^2 + 2Bx = C.$$

For the remainder of the section, we take this as our standard form for a quadratic equation. With this notation, we can appreciate the process of completing the square geometrically.

Algebraically, to make the left side of $x^2 + 2Bx = C$ a perfect square, we add B^2 to both sides, yielding

$$x^2 + 2Bx + B^2 = B^2 + C$$

or

$$(x + B)^2 = B^2 + C.$$

Taking square roots, we recover the quadratic formula, now in the form

$$x + B = \pm\sqrt{B^2 + C},$$

or

$$x = -B \pm \sqrt{B^2 + C}.$$

The change in our coefficients results in a quadratic formula without the cluttering appearances of 2 and 4.

Let's reconsider completing the square geometrically. Lengths can't be negative numbers. Let us assume, then, that B and C are positive, and in taking the square root of a positive number, we allow only the positive square root.

We begin with the equation

$$x^2 + 10x = 39$$

of Exercise 2.6, so that $B = 5$ and $C = 39$. Draw a square whose sides have an unknown length x and place rectangles atop it and to its right with sides of lengths 5 and x, as in Figure 2.1.

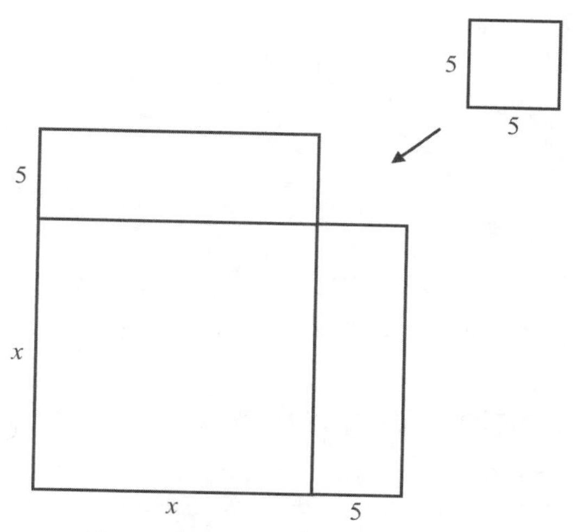

Figure 2.1. Completing the Square

Since the square has area x^2 and each rectangle has area $5x$, the figure drawn has area $x^2 + 10x$. We are given that

$$x^2 + 10x = 39,$$

so the area is also 39.

Looking at the figure, we can complete it by filling in the missing square in the upper right corner, which has side length 5. If we do fill it in—thereby completing the square—we get a square with side length $x + 5$. Its area is $(x+5)^2$. But also, since we are adding a 5×5 square to a figure that has area 39, we know that the area is $25 + 39$, or 64. Equating the two expressions for the area, we obtain

$$(x + 5)^2 = 64.$$

The larger square has side length given by $\sqrt{64}$, so the original square has side length x given by $-5+\sqrt{64}$, or 3. We have solved the quadratic equation geometrically, and we have seen that the algebraic process of completing the square has a geometric counterpart.

The general case is handled similarly. Given $x^2 + 2Bx = C$, with B and C arbitrary positive numbers, we draw the square whose sides have an unknown length x, then place rectangles atop it and to the right with sides of lengths B and x. The resulting figure has area $x^2 + 2Bx$, which we recognize as equal to C, thanks to the given equation. We complete the square by placing a square of side length B in the upper right corner, yielding a larger square with side length $x + B$. Its area is both $(x + B)^2$ and $B^2 + C$, yielding

$$(x + B)^2 = B^2 + C.$$

From this we obtain

$$x + B = \sqrt{B^2 + C}$$

or

$$x = -B + \sqrt{B^2 + C}.$$

We have derived the quadratic formula geometrically.

2.3 Changing Variables

Let us turn to our third approach for deriving the quadratic formula, the technique known as *change of variables*. We begin with a now-familiar example.

Exercise 2.8. Solve the equation

$$x^2 + 10x - 39 = 0$$

by changing variables.

(i) Let y related to x by $x = y - 5$. Substitute $y - 5$ for x in the equation and obtain an equation in y.

(ii) The new quadratic equation has the form $y^2 - d = 0$ for a constant d. We have a quadratic equation without a degree-one term.

(iii) Take square roots to obtain two values for y, then use $x = y - 5$ to obtain two solutions to the original quadratic equation.

We now take up the general case, with the goal of finding a change of variable that, as in Exercise 2.8, eliminates the degree-one term.

Exercise 2.9. Let $f(x) = x^2 + bx + c$. Let a be a real number. Introduce a new variable y by $x = y + a$ or $y = x - a$.

(i) Substitute $y + a$ for x in $f(x)$ and get a polynomial $g(y)$ in y. Determine it explicitly in terms of a, b, and c.

(ii) Examine $g(y)$ and determine a, in terms of b and c, so that $g(y)$ has the form $y^2 - d$. The number d will be expressible in terms of b and c.

(iii) With this a, the solutions to $g(y) = 0$ are the square roots of d.

(iv) Use the relationship between x and y to obtain that $x = a \pm \sqrt{d}$.

(v) Show that this is the quadratic formula.

We will use this technique in our treatment of cubic and quartic equations. It will allow us to drop the term of second-highest degree from the polynomial, as here we were able to drop the degree-one term.

2.4 A Discriminant

A quadratic polynomial $x^2 + bx + c$ may have two distinct real roots, one repeated real root, or no real roots, as we saw in Theorem 2.2. It is possible to determine which occurs from b and c, without finding the roots.

This is evident from the quadratic formula, since the roots, if they exist, are

$$-\frac{b}{2} \pm \frac{\sqrt{b^2 - 4c}}{2}.$$

We see that if $b^2 - 4c > 0$, there are two distinct real roots, if $b^2 - 4c = 0$, there is one multiple root, and if $b^2 - 4c < 0$, there are no real roots. We can obtain this independently of the roots by studying the shape of the graph of $y = x^2 + bx + c$.

To do so, we first need to review the shape of the graph of $y = x^2$. This was discussed in Section 1.6, where we saw that there is a turning point at $(0, 0)$, with the graph falling to $(0, 0)$ as x increases through negative values to 0 and rising as x increases through positive values, the shape being concave up throughout. (See Figure 2.2.)

The graph has an alternative description from the theory of conic sections. We won't be using it, but the issues are worth reviewing. The distance

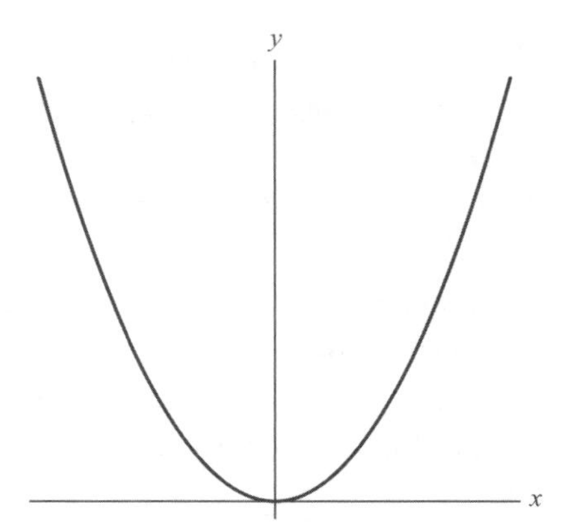

Figure 2.2. Graph of $y = x^2$

between two points (a, b) and (c, d) is the square root of $(c-a)^2+(d-b)^2$. (This is essentially the Pythagorean Theorem.) The distance from a point P to a line ℓ is defined to be the distance between P and the closest point Q on ℓ to P, with Q being the intersection of ℓ and the line perpendicular to ℓ that passes through P.

Exercise 2.10. Verify that the points in the plane equidistant from $(0, 1/4)$ and the line $y = -1/4$ are precisely those satisfying the equation $y = x^2$. (Hint: Set the squared distances equal to each other rather than the distances.)

In the theory of conic sections, a *parabola* is defined as the locus of points equidistant from a point, the parabola's *focus*, and a line not containing the point, the parabola's *directrix*. Using this language, Exercise 2.10 states that the graph of $y = x^2$ is the parabola with focus $(0, 1/4)$ and directrix $y = -1/4$.

The unique line that passes through a parabola's focus and meets the directrix at a right angle is the *axis of symmetry* for the parabola. The point where the axis of symmetry intersects the parabola is the parabola's *vertex*. For the graph of $y = x^2$, Exercise 2.10 shows that the focus is $(0, 1/4)$ and the directrix is $y = -1/4$. It follows that the axis of symmetry is the y-axis and the vertex is $(0, 0)$. (See Figure 2.3.)

The essential results about the behavior of the graph don't require calculus or the theory of conic sections, but do require results on the real numbers such as the intermediate value theorem. It lets us prove that x^2 assumes all positive real number values as x increases through positive real numbers,

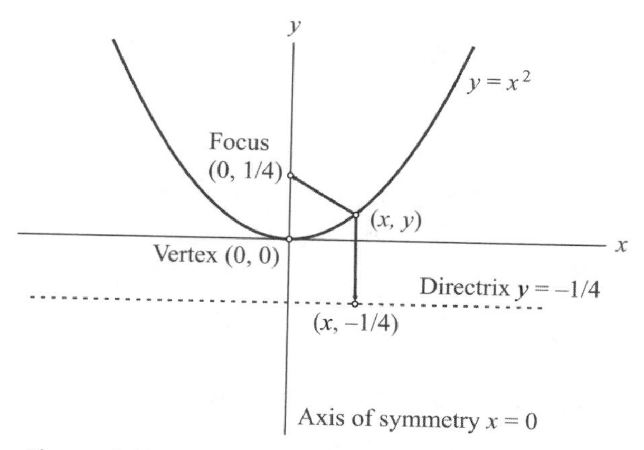

Figure 2.3. Focus and Directrix of the Parabola $y = x^2$

and similarly as x increases through all negative real numbers. Geometrically, this means the graph of $y = x^2$ goes through all positive heights as it falls to $(0, 0)$ on the left side of the y-axis and goes again through all positive heights as it rises on the right side of the y-axis.

Once we know the shape of the graph of $y = x^2$, we can use it for the graphs of other monic quadratic polynomials. Doing so depends on the following result, which tells us how to obtain equations for graphs that are obtained from known ones by horizontal or vertical translations.

Exercise 2.11. Suppose $y = f(x)$ is a function defined for all real number values of x.

(i) Given a real number C, verify that the graph of $y = f(x - C)$ is obtained from the graph of $y = f(x)$ by shifting every point to the right a distance of C.

(ii) Given a real number D, verify that the graph of $y = f(x) + D$ is obtained from the graph of $y = f(x)$ by shifting every point on it upward a distance of D.

(iii) Combine (i) and (ii) to obtain the result that the graph of $y = f(x - C) + D$ is obtained from the graph of $y = f(x)$ by shifting every point to the right a distance of C and upward a distance of D.

By combining Exercise 2.11 with the technique of completing the square, we can show that the graph of any monic quadratic polynomial is obtained from the graph of $y = x^2$ by vertical and horizontal translation.

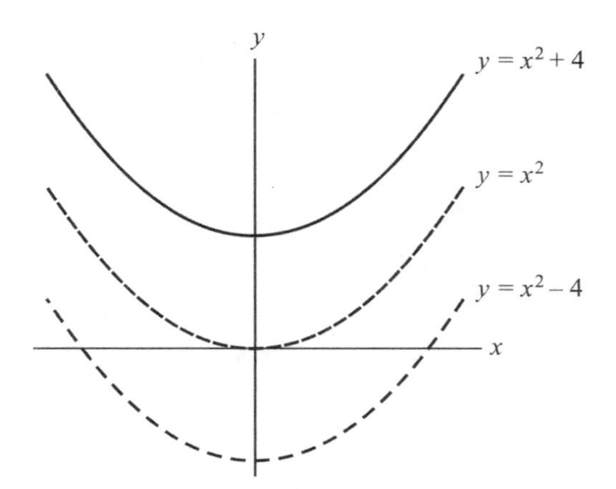

Figure 2.4. Graphs of three parabolas

Exercise 2.12. Let b and c be real numbers. By completing the square, we can write $y = x^2 + bx + c$ as

$$y = (x + \frac{b}{2})^2 + \left(c - \frac{b^2}{4}\right).$$

Deduce by Exercise 2.11 that the graph of $y = x^2 + bx + c$ is obtained from the graph of $y = x^2$ by a shift $-b/2$ to the right and $c - b^2/4$ upward.

Exercise 2.12 tells us that the graph of $y = x^2 + bx + c$ has the same shape as the graph of $y = x^2$, but is shifted up or down, left or right. Hence, the graph, like that of $y = x^2$, has a single turning point. This turning point may lie above, on, or below the x-axis, as illustrated in Figure 2.4 for x^2+4, x^2, and x^2-4. To decide which, we compute the y-coordinate of the turning point of $y = x^2 + bx + c$, which Exercise 2.12 allows us to do.

Exercise 2.13. Let b and c be real numbers. Using Exercise 2.12, conclude that the graph of $y = x^2 + bx + c$ has a local (and global) minimum at

$$(-b/2, c - b^2/4).$$

The graph of $x^2 + bx + c$ is the parabola with focus at $(-b/2, c - b^2/4 + 1/4)$, directrix $y = c - b^2/4 - 1/4$, axis of symmetry $x = -b/2$, and vertex at $(-b/2, c - b^2/4)$.

Now that we have found the turning point on the graph of a monic quadratic polynomial, we can use it to get information on the polynomial's roots.

Exercise 2.14. Using Exercise 2.13, show that the nature of the roots of $x^2 + bx + c$ is determined by the sign of $c - b^2/4$:

(i) If $c - b^2/4 < 0$, then the graph's turning point is below the x-axis. Conclude that the graph crosses the x-axis twice, so that $x^2 + bx + c$ has two distinct real roots.

(ii) If $c - b^2/4 = 0$, then the turning point is on the x-axis. Conclude that the graph of $x^2 + bx + c$ touches the x-axis once, so that $x^2 + bx + c$ has one real root. Since $x^2 + bx + c$ must factor as the product of two degree-one polynomials, deduce that the real root has multiplicity 2.

(iii) If $c - b^2/4 > 0$, then the turning point is above the x-axis. Conclude that the graph of $x^2 + bx + c$ does not touch or cross the x-axis, so that $x^2 + bx + c$ has no real roots.

(The three cases are illustrated in Figure 2.4.)

Exercise 2.14 shows that the nature of the roots of $x^2 + bx + c$ is determined by the quantity $c - b^2/4$. Any non-zero constant multiple of $c - b^2/4$ will determine their nature as well. Let's multiply by -4, obtaining the quantity $b^2 - 4c$. We will write δ for this and call it the *discriminant* of $x^2 + bx + c$. We see that it is the product of -4 and the y-coordinate of the turning point on the graph of $y = x^2 + bx + c$. Exercise 2.14 yields the result we are after:

Theorem 2.5. *Let $f(x) = x^2 + bx + c$ for real numbers b and c and let δ be the associated discriminant $b^2 - 4c$.*

(i) If $\delta > 0$, then $f(x)$ has two distinct real roots.

(ii) If $\delta = 0$, then $f(x)$ has a repeated real root.

(iii) If $\delta < 0$, then $f(x)$ has no real root.

2.5 History

In this section, we will look at the history of quadratic equations from ancient Babylonian civilization four millennia ago to Italy in 1500, touching occasionally on developments in algebra beyond the quadratic formula.

Every ancient civilization developed a body of mathematical knowledge, in part to serve the practical needs of measurement, construction, and commerce. The work of the ancient Greeks may be the most familiar, thanks to Euclid, but important contributions were made as well by Babylonian, Egyptian, Indian, and Chinese civilizations. Mathematical problems and solutions were often described through words, and geometry was used to express concepts that we might now address algebraically through notational systems not then available. General methods would be laid out implicitly through the working of a series of examples. All these civilizations addressed problems, in different guises, that we can now interpret as fitting under the heading of solving quadratic equations.

Let's begin with a look at some Babylonian work on problems reducible to quadratic equations. The first Babylonian dynasty, in the years ranging from around 1900 B.C.E. to 1600 B.C.E., made many contributions to human culture. Perhaps most notable is the code of Hammurabi, consisting of 282 laws that were written on clay tablets. Some of the earliest preserved mathematical texts date to this time. Babylonians used cuneiform, a wedge-like script that evolved from pictographs, for their writing. They also used base 60 in writing numbers, a choice that persists today in our methods for measuring time (minutes, seconds) and angle (degrees, minutes, seconds). Surviving on tablets are examples of the mathematical problems that Babylonians posed and solved. Many reduce to solving quadratic equations.

Our understanding of the mathematics on the cuneiform tablets owes much to the pioneering work of Otto Neugebauer, an Austrian scholar who studied mathematics at Munich and Göttingen in the 1920s. While at Göttingen, Neugebauer shifted his interest from mathematics to its history and did his doctoral research on Egyptian mathematics. He stayed on at Göttingen and began to study the Babylonian tablets that can be found at many of the major museums in Europe and the United States. His three-volume work *Mathematische Keilschrift-Texte* [44], published in 1935, translated and interpreted tablets found in museums from London and Paris to Berlin and Istanbul.

Neugebauer came to the United States in 1939 and continued his work, publishing *Mathematical Cuneiform Texts* [46] jointly with Abraham Sachs in 1945. He provided an overview of his findings in *The Exact Sciences in Antiquity* [45], published in 1952 based on a 1949 lecture series he presented at Cornell. These books are well worth a look, both for their content and for the tablet photographs.

In surveying the Babylonian treatment of algebra, Neugebauer highlighted its abstract nature, divorced from both geometric considerations and practical meaning:

> [G]eometrical concepts play a very secondary part in Babylonian algebra, however extensively a geometrical terminology may be used. It suffices to quote the existence of examples in which areas and lengths are added, or areas multiplied, thus excluding any geometrical inter-

pretation ... . Indeed, still more drastic examples can be quoted for the disregard of reality. We have many examples concerning wages to be paid for labor according to a given quota per man and day. Again, problems are set up involving sums, differences, products of these numbers and one does not hesitate to combine in this way the number of men and the number of days. It is a lucky accident that if the unknown number of workmen, found by solving a quadratic equation, is an integer. Obviously the algebraic relation is the only point of interest, exactly as it is irrelevant for our algebra what the letters may signify.

Neugebauer further explained [45, p. 42] that some of the Babylonian texts contain "collections of problems only, sometimes more than 200 on a single tablet of the size of a small printed page. These collections of problems are usually carefully arranged, beginning with very simple cases ... and expanding step by step to more complicated relations." Regarding one such series [45, p. 42], "one finds that they all have the same pair $x = 30, y = 20$ as solutions. This indicates that it was of no concern to the teacher that the result must have been known to the pupil. What he obviously had to learn was the method ... . From actually computed examples it becomes obvious that it was the general procedure, not the numerical result, which was considered important."

A tablet of particular interest is one catalogued in the British Museum as BM 13901. It originally contained twenty-four problems. Some are damaged, but as A. E. Berriman notes in his article "The Babylonian quadratic equation" [10, p. 185], "those that remain, however, are sufficient to reveal a carefully graduated course of instruction." Neugebauer studied the tablet in the third volume of *Mathematische Keilschrift-Texte* [44]. A discussion can also be found in the 2009 English translation of Jacques Sesiano's 1999 French work, *An Introduction to the History of Algebra: Solving Equations from Mesopotamian Times to the Renaissance* [59, pp. 10–16].

Let's look at the first problem of BM 13901. We will use the Babylonians' sexagesimal number system, as Neugebauer did in his translations. Commas separate the numbers in the 1 and 60 columns and a semi-colon separates the integer part from the decimal part, with the first position after the semi-colon being the number of sixtieths. For example, $1, 2$ is sexagesimal notation for the decimal number 62 and $0; 20$ is the fraction $20/60$, or $1/3$.

The problem is stated simply [59, p. 10]: "I added the area and the side of my square: $0; 45$." In standard fractional notation, this number is $45/60$, or $3/4$. The unknown is the length of the side of the square. Thus, we are being asked to solve the quadratic equation

$$x^2 + x = \frac{3}{4}.$$

What follows on the tablet is a recipe: "You put 1, the unit. You divide in two 1: 0; 30. You multiply by 0; 30: 0; 15. You add 0; 15 to 0; 45: 1. It is the square of 1. You subtract 0; 30, which you multiplied, from 1: 0; 30, the side of the square."

To make sense of this, it is perhaps better to think of the equation in the form $x^2 + bx + c = 0$, with $b = 1$ and $c = -3/4$. We are asked first to divide b by 2, yielding $b/2$. We are then told to square it, obtaining $b^2/4$. We add this to $-c$, obtaining $b^2/4 - c$. Finally we take the square root of this, $\sqrt{b^2/4 - c}$, and subtract the number we earlier multiplied, which was $b/2$. The final answer, then, is

$$-\frac{b}{2} + \sqrt{\frac{b^2}{4} - c},$$

or $1/2$. The recipe is the quadratic formula!

Let's take the same approach to the second problem of BM 13901 [59, p. 10]: "I subtracted from the area the side of my square: 14, 30." The sexagesimal number 14, 30, in decimal form, is $14 \times 60 + 30$, or 870. Thus, we are asked to solve the quadratic equation

$$x^2 - x = 870.$$

The solution is laid out as follows: "You put 1, the unit. You divide in two 1: 0; 30. You multiply by 0; 30: 0; 15. You add to 14, 30: 14, 30; 15. It is the square of 29; 30. You add 0; 30, which you multiplied, to 29; 30: 30, the side of the square."

Once again thinking of the equation as $x^2 + bx + c = 0$, this time with $b = -1$ and $c = -870$, we can describe the solution as follows: We first form the quantity $-b/2$. We then square it and add the result to $-c$ to get $b^2/4 - c$. The next step is to determine the square root, to which we add the number we earlier used for multiplying, which was $-b/2$. The result—30—is the desired answer, and once again it is what we obtain by using the quadratic formula.

It turns out that the two problems are not typical. Few examples of single quadratic equations have been found. More common are problems involving two variables and two equations, one equation having the form $xy = 1$, the other being a linear equation in x and y. They can be reduced to solving a single quadratic equation.

Consider for instance Problem-Text T in Neugebauer and Sachs's *Mathematical Cuneiform Texts*. They open their commentary [46, p. 116] with the observation that the text "can best be compared to an extensive collection of problems from a chapter of a textbook. It is obvious that a collection of this sort was used in teaching mathematical methods." The student is asked to reduce each problem to what Neugebauer and Sachs refer to [46, p. 117] as "the 'normal' form of quadratic equations which gives the product of the unknown quantities and their sum or difference. ... The whole system of a main example with all its variants serves the purpose of giving the general rule of solution; this corresponds in a certain sense to an algebraic formula in which the letters can be replaced by special numbers in each individual case."

The normal form is the problem we studied in Section 2.1. A representative example is the first problem on tablet AO 8862 at the Louvre in Paris, treated by Neugebauer in the first volume of *Mathematische Keilschrift-Texte* [44, pp. 108–123] and by Sesiano as well [59, p. 12]. Van der Waerden opens the chapter on Babylonian mathematics in *Science Awakening* with this example [64, pp. 63–65]: "Length, width. I have multiplied length and width, thus obtaining the area. Then I added to the area, the excess of the length over the width: 3, 3 (i.e., 183 was the result). Moreover, I have added length and width: 27. Required length, width and area."

We are asked to solve the two equations

$$xy + (x - y) = 183$$

and

$$x + y = 27.$$

The tablet then lays out a sequence of numerical calculations that lead to the answer. They can be interpreted as instructing the student to add the equations, yielding $x(y + 2) = 210$, and to add 2 to both sides of the second equation, yielding $x + (y + 2) = 29$. This transforms the original equations to normal form, x and $y + 2$ being two numbers whose product is 210 and whose sum is 29.

It is important to note that scholarship on Babylonian mathematics in recent decades has yielded a perspective sharply different from that engendered by Neugebauer's work. Jens Høyrup, in his 2002 study *Lengths, Widths, Surfaces: A Portrait of Old Babylonian Algebra and Its Kin* [35], observes that

> The Babylonian algebra which most historians of mathematics found in Neugebauer's works looked astonishingly modern and similar to

> ours. It is the purpose of the present book to replace this standard interpretation by a less modernizing reading.... The mathematical texts are school texts. They contain no theorems and no theoretical investigations.... [Their authors] were teachers of computation, at times teachers of pure, unapplicable computation ... but they remained teachers, teachers of scribe school students who were later to end up applying mathematics to engineering, managerial, accounting, or notarial tasks.

Høyrup's own analysis of the first problem on AO 8862 #1 [35, pp. 169–170] emphasizes its concrete nature: "The text starts by stating that a rectangular surface or field is built, that is, marked out; after pacing off its dimensions, the speaker 'appends' the excess of the length over the width to it; the outcome is 3, 3. Even this is done quite concretely in the terrain. Then he 'turns back' and reports the accumulation of the length and the width to be 27."

In *Mathematics in Ancient Iraq: A Social History* [56], Eleanor Robson offers an enlightening side-by-side comparison of van der Waerden's and Høyrup's analyses [56, pp. 276–278]. "In van der Waerden's translation of 1954 the problem is entirely numerical ... In his reading 'the Babylonians' are just like modern mathematicians: they use 'symbols' and 'equations,' which means that the problem can 'safely' be expressed as modern algebra. Høyrup, by contrast, opens his comments on the same problem with an interpretative diagram that does not appear on the cuneiform tablet and continues [as quoted above]. All of van der Waerden's apparently arithmetical numbers turn out to have dimension: they are particular lengths and areas that are manipulated physically."

Another common form taken by problems on the tablets is to find two numbers from known values of the sum of their squares and of their sum (or difference). An example is problem 9 of BM 13901 [59, p. 15]: "I added the area of my two squares: 1300. The side of one exceeds the side of the other by 10." Here we are asked to solve the equations

$$x^2 + y^2 = 1300$$

and

$$x - y = 10.$$

This is easily converted to a quadratic equation in one variable. In reinterpreting the problem this way, we are placing a modern overlay on the data of the tablet. Høyrup offers his own analysis [35, pp. 66–70], attempting to interpret the tablet's prescription as a series of instructions for tearing out, inscribing, and appending squares of given dimensions.

We have given more attention to Babylonian mathematics and the implicit treatment within it of quadratic equations than might seem necessary. However, it is difficult to avoid the temptation. As Eleanor Robson concludes [56, p. 290]:

> Compared to the difficulties of grappling with fragmentary and meagre nth-generation sources from other ancient cultures the cuneiform evidence is concrete, immediate, and richly contextualised. ... This opens a unique window onto the material, social, and intellectual world of the mathematics of ancient Iraq that historians of other ancient cultures can only dream of.

Around 600 B.C.E., a new mathematical current arose within the Greek civilization of the Mediterranean, starting with the work of Thales and Pythagoras. The Greek tradition would have enormous influence on the development of mathematics, and much has been written about it. Perhaps of greatest importance was the introduction of the axiomatic method and deductive proof. For a concise introduction, William Berlinghoff and Fernando Gouvêa have an informative survey [9, pp. 14–24]. (See S. Cuomo's *Ancient Mathematics* [15] for a more detailed study.) The most famous of the Greek mathematicians is Euclid, who lived around 300 B.C.E. in Alexandria. Little is known about him, or about the origins of his greatest work, the *Elements*. For instance, no definitive answer can be given to the question of how much of the *Elements* is due to him and how much is a compendium of earlier work.

The *Elements* provided the model for the axiomatic method and laid the basis for geometry for centuries. Most of its thirteen books are devoted to geometry, with plane geometry treated first and regular polyhedra in three-space covered in the concluding Book 13. Books 7 to 9 contain some of the most famous results of elementary number theory, such as Euclid's proof that there are infinitely many prime numbers. Book 2 provides solutions to geometric questions about area that amount to solving simultaneous equations $x + y = a$ and $xy = b$ for constants a and b [65, pp. 77–80]. As we know, this reduces to solving a quadratic equation.

Jumping ahead five centuries to the late stages of the Greek tradition, we come to Diophantus, another Alexandrian. Diophantus wrote *Arithmetica*, a collection of about 200 problems and their solutions. *Arithmetica* consisted of thirteen chapters, of which six were preserved in Greek and four in an Arabic translation from the ninth century, but the Arabic chapters were essentially lost until their rediscovery in 1968.

In contrast to Euclid's geometric algebra, Diophantus works in a manner we would recognize as more strictly algebraic, with symbols for the lower positive and negative powers of an unknown, rules of multiplication, and rules of substitution. Many of his problems reduced to equations in two unknowns. Among them were problems studied by the Babylonians, such as finding two numbers whose sum and product are given. Solutions were to be found in rational numbers.

The Greek chapters of *Arithmetica* had little influence until a revival of interest in them among Italian mathematicians late in the sixteenth century and even more so by the French mathematician Pierre de Fermat in the seventeenth century. In Book II, Diophantus poses the problem to divide a square into two squares, that is, to find solutions to $x^2 + y^2 = z^2$. Although we have not discussed this problem, it has a history as long as that of the quadratic equation. All ancient civilizations studied it, including two we have not discussed, those of Egypt and China.

Fermat owned a copy of a 1621 Latin translation of *Arithmetica* made by the French mathematician and scholar Claude-Gaspard Bachet (1581–1638). (Bachet's best known work is *Problèmes plaisants et délectables qui se font par les nombres* [6], his 1612 collection of arithmetic puzzles and tricks that was the basis for many subsequent books on recreational mathematics.) Some time around 1637, Fermat wrote the famous marginal note that for integers n greater than 2, the equation $x^n + y^n = z^n$ has no rational solution, adding the tantalizing remark that he had a "truly marvelous proof" that his margin was too narrow to contain.

Whether Fermat's statement was true would remain one of the great open questions in mathematics until it was settled in the affirmative in 1995 by Andrew Wiles and Richard Taylor. In the meantime, Diophantus became the eponym for an entire branch of mathematics, Diophantine analysis, which brings number-theoretic and geometric methods to bear on the question of whether algebraic equations in several unknowns have integer or rational solutions. (See [7], [8, pp. 35–57], and [59, pp. 31–51] for more information on Diophantus's work.)

In tracing the development of algebra, let us move east to India, where the mathematician and astronomer Aryabhata lived from 476 C.E. to about 550 C.E. In his work *Aryabhatiya*, he solves some problems that reduce to quadratic equations. Satya Prakash observes in a book on the later Indian astronomer-mathematician Brahmagupta [53, p. 215] that Aryabhata has "given the solutions of a few quadratic equations, but he nowhere gives the procedure of solving these equations."

Brahmagupta, who lived from around 598 to 665, wrote several major works, including *Brahmasphutasiddhanta*. Some of his work on arithmetic and algebra was translated into English in 1817 by H.T. Colebrooke

[14]. Here is one problem Brahmagupta poses [14, p. 346]: "When does the residue of revolutions of the sun, less one ... equal to the square root of two less than the residue of revolutions, less one, multiplied by ten and augmented by two?" Translating into modern algebraic notation, Brahmagupta is looking for a solution of the equation $x^2 - 86x = 249$. Following Colebrooke in his translation, let us call 249 the "absolute number" and x the "middle term." With this terminology, Brahmagupta's proposed solution amounts to a statement of the quadratic formula:

> Now, from the absolute number, multiplied by four times the coefficient of the square, and added to the square of the coefficient of the middle term, the square root being extracted, and lessened by the coefficient of the middle term, the remainder is divided by twice the coefficient of the square, yields the value of the middle term.

Satya Prakash comments [53, p. 215] that Brahmagupta "undoubtedly is not the discoverer of these rules, but perhaps for the first time in the history of algebra we find the process of solving a quadratic equation so clearly indicated."

Brahmagupta's greatest contribution to algebra is his study of integer solutions to the equation $x^2 - ny^2 = \pm 1$, where n is a positive integer that is not a square. Such an equation is now known as *Pell's equation* and arises naturally in a variety of contexts. The twelfth century Indian mathematician Bhaskara obtained more definitive results. More information on this work can be found in V. S. Varadarajan's *Algebra in Ancient and Modern Times* [67], which has much to say about other cultures as well. See too B. L. van der Waerden's *Geometry and Algebra in Ancient Civilizations* [65].

The subsequent centuries were ones of exciting intellectual developments in the Islamic world. Most notable for us is the work of Muhammad ben Musa al-Khwarizmi, a Persian who lived roughly between 780 and 850 near Baghdad. Al-Khwarizmi contributed to many fields, including mathematics, astronomy, and geography. Perhaps his greatest work is his treatise on algebra, *The Compendious Book on Calculation by Completion and Balancing*, written in Arabic around 825. In it, he systematically studied linear and quadratic equations using the techniques of *al-jabr* and *al-muqabala.* The process of *al-jabr* is that of adding equal terms to both sides of an equation in order to eliminate negative terms. In contrast, *al-muqabala* is the reduction of positive terms by subtracting equal amounts from both sides of an equation.

Al-Khwarizmi's work would have a significant influence on European mathematics several centuries later, thanks to its translation into Latin in the twelfth century by Robert of Chester and Gerard of Cremona and in the thirteenth century by William of Luna. Robert of Chester's version, *Liber algebrae et almucabola*, was translated into English in 1915 by Louis Charles Karpinski, who also provided notes and an introduction [36]. Karpinski's edition is well worth a look, as is the more recent *Al-Khwarizmi: The Beginnings of Algebra* [54] by Roshdi Rashed, which includes a translation (with the original Arabic text on facing pages) and commentary. Rashed also writes about the traditions of calculation in the eighth century and al-Khwarizmi's knowledge of Greek and Indian mathematical literature.

At the outset of his book, al-Khwarizmi introduces three types of quantities: roots, squares, and numbers, the root being what we would label x, the square being x^2, and number being number. He then gets right to business, classifying linear and quadratic equations into six forms and illustrating how to solve each with examples. We would regard all six types as one, but since al-Khwarizmi does not use negative numbers or 0, he is obliged to consider separate cases, with a, b, and c always positive:

(1) Squares equal roots, or what we would describe as $ax^2 = bx$.

(2) Squares equal numbers, or $ax^2 = c$.

(3) Roots equal numbers, or $bx = c$.

(4) Squares and roots equal numbers, or $ax^2 + bx = c$.

(5) Squares and numbers equal roots, or $ax^2 + c = bx$.

(6) Roots and numbers equal squares, or $bx + c = ax^2$.

(In addition to [36] and [54], see also the discussion of al-Khwarizmi's work in *The Beginnings and Evolution of Algebra* by I.G. Bashmakova and G.S. Smirnova, from which the above summary is drawn [8, p. 50].)

What is novel about the opening of al-Khwarizmi's book is that in it he lays out the mathematical issues in purely algebraic terms before turning to the examples—drawn from commerce and inheritances—that fill the later pages of the book. The focus is on equations in the abstract, classified by degree. Rashed explains in the introductory essay to his translation [54, p. 24] that what al-Khwarizmi

> does cannot be reduced to anything to be found in other traditions, such as those of the Babylonians, of Diophantus, of Heron of Alexandria, of Aryabhata or of Brahmagupta. It is not in the course of solving problems that al-Khwārizmī finds these equations. The classification in fact precedes the problems. It is introduced deliberately as the necessary first step in the construction of a theory of equations of the first

and second degree; and this theory will become the nucleus of a mathematical discipline.

The first of al-Khwarizmi's six cases that interests us is (4), and the example he uses to illustrate it is none other than the quadratic equation $x^2 + 10x = 39$ that we studied in Sections 2.2. This equation is not distinguished in any mathematical sense, but since it is the first one for which al-Khwarizmi employs the technique of completing the square, it has resonated through history. Or, as Karpinski suggests [36, p. 19], the equation "runs like a thread of gold through the algebras for several centuries, appearing in the algebras of the three writers mentioned, Abu Kamil, al-Karkhi and Omar al-Khayyami, and frequently in the works of Christian writers, centuries later." Here is al-Khwarizmi's own description of the example [54, p. 100]:

> "Squares plus roots are equal to a number", as when you say: a square plus ten roots are equal to thirty-nine dirhams [a unit of measure]; namely, if you add to any square [a quantity] equal to ten of its roots, the total will be thirty-nine.
>
> Procedure: you halve the number of the roots which, in this problem, yields five; you multiply it by itself; the result is twenty-five; you add it to thirty-nine; the result is sixty-four; you take the root, that is eight, from which you subtract half the number of the roots, which is five. The remainder is three, that is the root of the square you want, and the square is nine.

Upon completing his treatment of the remaining types of quadratic equation, al-Khwarizmi returns to the equation $x^2 + 10x = 39$ and illustrates two ways of solving it geometrically. The second is the one we used in Section 2.2.

Rashed concludes his introductory essay by turning to the question of Indian influence on al-Khwarizmi. Two centuries ago, in translating the work of the Indians, Colebrooke suggested [14, pp. xx–xxi] that al-Khwarizmi was a mere borrower. This assessment is misguided, but the extent to which Islamic mathematicians and astronomers were familiar with and influenced by Indian literature is an interesting question. Rashed emphasizes what is original to al-Khwarizmi [54, p. 77]:

> Comparing the two texts, that of Brahmagupta and that of al-Khwarizmi, reveals irreducible differences. ... Brahmagupta arrived at the quadratic equation in one unknown in the course of solving a problem in astronomy. In other words, he did not give himself the equation as

such with a view to solving it. This link between problem and equation, which is found in other mathematics, this as it were empirical grounding for the equation, has vanished in al-Khwarizmi's programme. From the start, al-Khwarizmi proceeded by defining basic terms, which were then combined to give him the ideal canonical equations with which his theory is concerned. This new approach breaks the close link between problems and equations. As for problems, al-Khwarizmi turns to them later, as exercises in algebra, that is as providing an area in which he can apply the theory of equations that he has already constructed. ...

... al-Khwarizmi wanted to construct "a form of calculation" for unknowns that was independent of what they represented, that is a new mathematical discipline, that by its very nature was subject to the rules of proof. There is no trace of any such programme in the work of his predecessors.

Al-Khwarizmi was one of several Islamic scholars who made important contributions to algebra. Others include Thabit ibn Qurra, who spent much of his career in Baghdad, studying medicine, philosophy, mathematics, and astronomy as well as translating Euclid and other Greeks into Arabic. He died in 901. Also worth mentioning is the famed Persian poet, philosopher, and mathematician Omar Khayyam, who studied in Samarkand and worked in Bukhara (both in modern-day Uzbekistan), dying in 1131. He wrote a book on algebra, studying quadratic equations from a geometric viewpoint as in Euclid. The work of Thabit and Omar Khayyam is discussed in detail in van der Waerden's *A History of Algebra From Al-Khwarizmi to Emmy Noether* [66, pp. 15–31].

While mathematics was flourishing in Arabic, Persian, and Indian lands, the medieval era in Europe was mathematically more quiescent. The revival of significant European mathematical activity occurred first in Italy, perhaps because its cities were major trade centers with connections to Arabic ports along the Mediterranean. The great Italian mathematician Leonardo da Pisa—better known to us as Fibonacci—provides testimony to this effect in the prologue to his influential 1202 work *Liber Abaci* [28, pp. 15–16], or the *Book of Calculation*:

As my father was a public official away from our homeland in the Bugia [the Algerian port city of Bejaia] customshouse established for the Pisan merchants who frequently gathered there, he had me in my youth brought to him, looking to find for me a useful and comfortable future; there he wanted me to be in the study of mathematics and to be taught for some days. There from a marvelous instruction in the

> art of the nine Indian figures, the introduction and knowledge of the art pleased me so much above all else, and I learnt from them, whoever was learned in it, from nearby Egypt, Syria, Greece, Sicily and Provence, and their various methods, to which locations of business I travelled considerably afterwards for much study, and I learnt from the assembled disputations.

Fibonacci is best known for the sequence of numbers that bears his name, the *Fibonacci numbers*, which arise in his solution to the following problem [28, pp. 404–405]:

> A certain man had one pair of rabbits together in a certain enclosed place, and one wishes to know how many are created from the pair in one year when it is the nature of them in a single month to bear another pair, and in the second month those born to bear also.

This is sufficiently famous that we will take a moment to review Fibonacci's conclusion. He works out the numbers month by month, then explains that "we added the first number to the second, namely the 1 to the 2, and the second to the third, and the third to the fourth, and the fourth to the fifth, and thus one after another until we added the tenth to the eleventh, namely the 144 to the 233, and we had the above written sum of rabbits, namely 377, and thus you can in order find it for an unending number of months."

In Chapter 15 of *Liber Abaci*, Fibonacci examines the algebra of quadratic equations, following al-Khwarizmi's treatment closely. For example [28, p. 554], Fibonacci introduces "roots, squares, and simple numbers" and explains that "in the solutions of problems there are six modes of which three are simple, and three are composite," just as in al-Khwarizmi's classification of quadratic equations. Of particular importance is Fibonacci's use of Hindu-Arabic numerals, which led to their adoption in Europe.

For a brief survey of the algebraic work of Italian mathematicians in the subsequent three centuries, concluding with Luca Pacioli (1445–1514), one can turn to [66, pp. 42–47]. Pacioli provided an overview of the mathematical knowledge of the time in his 1494 book *Summa de arithmetica, geometrica, proportioni e proportionalità* [50], published in Venice. The most influential material in this book would turn out to be his account of double-entry bookkeeping, thanks to which he has been called the Father of Accounting. We will return to Pacioli for a moment, in Section 3.5, to begin our survey of sixteenth-century Italian contributions to the solving of cubic equations.

Let us close this historical discussion with an excerpt from Karpinski's introduction [36, pp. 11–12] to his translation of al-Khwarizmi, in which he describes the study of quadratic equations as an inevitable stage in the history of human thought.

> Algebra is not, as often assumed, an artificial effort of human ingenuity, but rather the natural expression of man's interest in the numerical side of the universe of thought. Tables of square and cubic numbers in Babylon; geometric progressions, involving the idea of powers, together with linear and quadratic equations in Egypt; the so-called Pythagorean theorem in India, and possibly in China, before the time of Pythagoras; and the geometrical solution of quadratic equations even before Euclid in Greece, are not isolated facts of the history of mathematics. While they do indeed mark stages in the development of pure mathematics, this is only a small part of their significance. More vital is the implication that the algebraical side of mathematics has an intrinsic interest for the human mind not conditioned upon time or place, but dependent simply upon the development of the reasoning faculty. We may say that the study of powers of numbers, and the related study of quadratic equations, were an evolution out of a natural interest in numbers; the facts which we have presented are traces of the process of this evolution.

3

Cubic Polynomials

In this chapter, we will take our first look at cubic equations and the famous formula for their solution known as Cardano's formula. Girolamo Cardano, for whom the formula is named, was a sixteenth-century Italian scholar. The story of the formula's discovery is complex, as we will see in Section 3.5, and credit must be shared with Scipione del Ferro and Niccolò Fontana.

Our results in this chapter will be imprecise, because we are lacking what turns out to be an essential tool: complex numbers. However, it is this first look that will reveal the need for complex numbers. After developing the basic facts about them in Chapter 4, we will return to cubic equations in Chapter 5 and treat them with appropriate care.

3.1 Reduced Cubics

Quadratic polynomials need not have roots. In contrast, a cubic polynomial must have a root. This is a special case of Theorem 1.14, which stated that any odd degree polynomial has a root. (Alternatively, once we have Cardano's formula and complex numbers, the existence of a root will follow automatically from algebra alone.) Let us record this as a theorem.

Theorem 3.1. *Let $f(x)$ be a cubic polynomial. Then there is a real number a such that $f(a) = 0$.*

Theorem 1.4 states that, given a polynomial $f(x)$ and a real number a, if $f(a) = 0$, then $x - a$ divides $f(x)$. Combining this with Theorem 3.1, we find that any cubic polynomial can be factored as a product of linear and quadratic polynomials.

Exercise 3.1. Suppose $f(x)$ is a monic cubic polynomial.

(i) Use Theorems 1.4 and 3.1 to deduce that $f(x)$ factors as

$$(x-a)(x^2+mx+n)$$

for some real numbers a, m, and n.

(ii) Deduce that any real root of $f(x)$ besides a is a root of x^2+mx+n. (Hint: Suppose r is a root and $r \neq a$. Substitute r in $f(x)$, using the factorization in (i). What do you get?)

(iii) Apply Theorem 2.2 to x^2+mx+n to prove Theorem 3.2.

Theorem 3.2. *Let $f(x)$ be a monic, cubic polynomial. Exactly one of the following occurs.*

(i) $f(x)$ has one real root a, of multiplicity 3, and factors as

$$(x-a)^3.$$

(ii) $f(x)$ has two distinct real roots a_1 and a_2, of multiplicities 1 and 2, and factors as

$$(x-a_1)(x-a_2)^2.$$

(iii) $f(x)$ has three distinct simple real roots a_1, a_2, and a_3, and factors as

$$(x-a_1)(x-a_2)(x-a_3).$$

(iv) $f(x)$ has one simple real root a and factors as

$$(x-a)s(x),$$

where $s(x)$ is a quadratic polynomial that does not factor as a product of linear polynomials.

A cubic polynomial of the form

$$y^3+py+q$$

is called a *reduced* cubic. In the next exercise, we will see that a change of variable allows us to pass from an arbitrary cubic polynomial to a reduced one. This is analogous to the change of variable used in Exercise 2.9 to pass from a quadratic polynomial to a new one without a degree-one term, and will simplify the task of solving cubic equations.

Exercise 3.2. Let

$$f(x) = x^3 + bx^2 + cx + d,$$

for real numbers b, c, and d. Given another real number a, let's see what happens under the change of variable $x = y + a$, or $y = x - a$.

(i) Substitute $y+a$ for x and obtain a polynomial $g(y)$ in the new variable y. Write it as

$$y^3 + By^2 + Cy + D$$

and obtain formulas for the coefficients B, C, and D in terms of a and the old coefficients b, c, and d.

(ii) Observe that there is a choice of a for which $B = 0$. Thus for this a, changing variables provides a new polynomial $g(y)$ that is reduced.

(iii) Verify that $g(y)$ is the reduced polynomial described in Theorem 3.3.

(iv) What is the relation between a root of $f(x)$ and a root of $g(y)$? In particular, given a root s of $g(y)$, describe a root r of $f(x)$.

(v) Conclude that solving $g(y) = 0$ allows us to solve $f(x) = 0$.

Theorem 3.3. *Let b, c, and d be real numbers. Given a cubic polynomial $x^3 + bx^2 + cx + d$, the change of variable $x = y - b/3$ yields the reduced cubic polynomial*

$$y^3 + \left(-\frac{b^2}{3} + c\right) y + \left(\frac{2b^3}{27} - \frac{bc}{3} + d\right).$$

We have shown that we can pass from the problem of solving an arbitrary cubic equation to the equivalent problem of solving a cubic equation of the form

$$y^3 + py + q = 0,$$

and that being able to solve equations of this simpler type allows us to solve arbitrary cubic equations.

Exercise 3.3. Suppose we wish to solve the cubic equation

$$x^3 - 3x^2 - 4x + 12 = 0.$$

(i) What reduced cubic equation would we solve in order to find the solutions?

(ii) How are the solutions of the reduced cubic equation related to the solutions of the original cubic equation?

In Section 3.2, we will solve the reduced cubic equation in Exercise 3.3, allowing us to find the solutions of the original cubic equation.

There are two families of reduced cubic equations that take on an even simpler form, those for which $p = 0$ and those for which $q = 0$. Let's discuss these.

Exercise 3.4. Consider the cubic equation $y^3 + py = 0$.

(i) Observe that $y = 0$ is a solution and that the other solutions are solutions to $y^2 + p = 0$.

(ii) Deduce that if $p > 0$ then $y = 0$ is the only solution and 0 is a simple root of $y^3 + py$; if $p = 0$ then $y = 0$ is the only solution and is a repeated root of $y^3 + py$; and if $p < 0$ then there are three distinct solutions: $y = 0$ and $y = \pm\sqrt{-p}$.

Exercise 3.5. Consider the cubic equation $y^3 + q = 0$. If $q = 0$, we are studying the equation $y^3 = 0$. Obviously the only solution is 0, which has multiplicity 3 as a root of y^3. Assume for the remainder of this exercise that $q \neq 0$.

(i) By Theorem 3.1, the equation must have a solution, and any solution is a cube root of $-q$.

(ii) Suppose r is such a cube root: $r^3 = -q$. Thus, we can rewrite the polynomial $y^3 + q$ as $y^3 - r^3$. Show that

$$y^3 - r^3 = (y - r)(y^2 + ry + r^2).$$

(iii) Show that $y^2 + ry + r^2$ has no real roots. (What is its discriminant?)

(iv) Conclude that r is the lone root of $y^3 + q$ and that r has multiplicity 1.

3.2 Cardano's Formula

We have seen in Section 3.1 that every cubic polynomial has at least one root and that by change of variable we can reduce the problem of finding a root of a given cubic polynomial $x^3 + bx^2 + cx + d$ to the problem of finding a root of a cubic polynomial of the form $y^3 + py + q$. What would be a suitable solution to the reduced problem?

At the end of Section 2.1, we observed that the quadratic formula does not so much solve quadratic equations as provide a reduction procedure, one that uses algebra to reduce the problem of solving a quadratic equation to the problem of calculating square roots of real numbers. Algebra offers no guidance on finding the square roots. Similarly, in solving cubic equations, we should be content if we reduce the problem to one of calculating square and cube roots. The calculation of cube roots of real numbers, like the calculation of square roots, is not a problem we should expect algebra to solve. In fact—and this is an essential part of our story—we will discover that we must be content with something less: a reduction to square root calculations of real numbers and cube root calculations of entities more general than real numbers.

No point delaying. Here's the formula—*Cardano's formula*—for a solution of the reduced cubic equation $y^3 + py + q = 0$:

$$y = \sqrt[3]{-\frac{q}{2} + \sqrt{\frac{p^3}{27} + \frac{q^2}{4}}} + \sqrt[3]{-\frac{q}{2} - \sqrt{\frac{p^3}{27} + \frac{q^2}{4}}}$$

We will explore its history in Section 3.5. For now, let us just note that Cardano did not write down such a formula explicitly. Rather, he illustrated how to solve reduced cubic equations through examples.

Exercise 3.6. Substitute the expression for y given by Cardano's formula into the polynomial y^3+py+q and verify, after expanding and simplifying, that it equals 0.

Cardano's formula expresses the solution to $y^3 + py + q = 0$ in terms of the coefficients p and q, addition, multiplication, division by some constants, and the taking of square and cube roots. Thus, just as the quadratic formula reduces the solution of quadratic equations to square root calculations, Cardano's formula reduces the solution of cubic equations to square root and cube root calculations. Once we use Cardano's formula to find a real root r of $y^3 + py + q$, we can divide $y^3 + py + q$ by $y - r$ to obtain a quadratic polynomial and use the quadratic formula to find any additional real roots.

The formula has one worrisome feature, the appearance of the square root of $p^3/27 + q^2/4$. In the quadratic formula, whenever the quadratic polynomial being studied has a real root, the quantity whose square root must be calculated is non-negative. Cubic polynomials, in contrast to quadratics, always have real roots. If the value of $p^3/27+q^2/4$ for a reduced cubic polynomial is negative, we may be in for trouble in attempting to use Cardano's formula. We'll soon see what issues arise.

Before using Cardano's formula in specific examples, let's see how we might derive it. There are many ways. In the next exercise, we will work through an approach described by François Viète in 1591. More will be said about Viète in the historical discussion of Section 5.8.

Exercise 3.7. We begin with the cubic polynomial $y^3 + py + q$. The key idea (by no means obvious!) is to introduce a new variable z satisfying

$$y = z - \frac{p}{3z}.$$

If $p = 0$, this change of variable accomplishes nothing, so let's assume until the last part of the exercise that $p \neq 0$.

(i) Substitute $z - p/3z$ for y in $y^3 + py + q = 0$, expand the cubed term, simplify, and obtain an equation in z:

$$z^3 - \frac{p^3}{27z^3} + q = 0.$$

(ii) Multiply by z^3 to clear denominators and obtain

$$z^6 + qz^3 - \frac{p^3}{27} = 0.$$

(iii) This is a quadratic equation in z^3. Use the quadratic formula to obtain

$$z^3 = -\frac{q}{2} \pm \frac{\sqrt{q^2 + (4p^3/27)}}{2}.$$

(iv) Introduce R as an abbreviation for $p^3/27 + q^2/4$ and rewrite the last equality as

$$z^3 = -\frac{q}{2} \pm \sqrt{R}.$$

(v) There are two possible values for z^3, either $-(q/2)+\sqrt{R}$ or $-(q/2)-\sqrt{R}$. Multiply them and simplify to get

$$\left(-\frac{q}{2} + \sqrt{R}\right)\left(-\frac{q}{2} - \sqrt{R}\right) = \left(-\frac{p}{3}\right)^3.$$

(vi) Take cube roots of both sides and deduce that the two values of z have a product satisfying

$$\left(\sqrt[3]{-\frac{q}{2}+\sqrt{R}}\right)\left(\sqrt[3]{-\frac{q}{2}-\sqrt{R}}\right)=-\frac{p}{3}.$$

(vii) This means that if z is the cube root of $-(q/2)+\sqrt{R}$, then $-p/3z$ is the cube root of $-(q/2)-\sqrt{R}$.

(viii) Recall that z was introduced to satisfy $y = z - p/3z$. The two terms on the right of this equation, z and $-p/3z$, are the cube roots of $-(q/2)+\sqrt{R}$ and of $-(q/2)-\sqrt{R}$.

(ix) Conclude that y is the sum of the two cube roots:

$$y=\sqrt[3]{-\frac{q}{2}+\sqrt{R}}+\sqrt[3]{-\frac{q}{2}-\sqrt{R}}.$$

(x) We have assumed that $p \neq 0$. If $p = 0$, then the equation we are solving is $y^3 + q = 0$ and the formula still holds. (Check that one summand becomes 0 and the other is the cube root of $-q$.)

We have obtained Cardano's formula! That wasn't hard at all. However, in the derivation of the formula, we have not been precise on certain points. The imprecisions will be addressed in Exercise 5.1, where we review the derivation with the benefit of complex numbers. For now, let us ignore these imprecisions and give the formula a try.

Exercise 3.8. Solve $y^3 - 3y + 2 = 0$.

(i) Use Cardano's formula to find one solution r.

(ii) Use it to factor $y^3 - 3y + 2$ as the product of linear and quadratic polynomials and find the other two solutions.

The next equation is a famous example of Cardano.

Exercise 3.9. Solve $y^3 + 6y - 20 = 0$.

(i) Use Cardano's formula and obtain

$$y=\sqrt[3]{10+6\sqrt{3}}+\sqrt[3]{10-6\sqrt{3}}.$$

(ii) Observe that the function $y^3 + 6y - 20$ is always rising as y increases. (Consider the behavior of the separate terms y^3 and $6y$.)

(iii) Deduce that the graph of $y^3+6y-20$ crosses the y-axis (the horizontal axis) only once, so that $y^3 + 6y - 20 = 0$ has only one real solution.

(iv) Conclude that the expression we have obtained for y using Cardano's formula must be the lone real solution.

We can stop here, content that we have found the single real root of $y^3+6y-20$. Or we can use a calculator to determine at least approximately what the sum of the two cube roots is. If we do this, we discover that it is close to 2. Substituting 2 for y in $y^3 + 6y - 20$ to see how close to 0 we are, we further discover that

$$2^3 + 6 \cdot 2 - 20 = 8 + 12 - 20 = 0.$$

In other words, 2 is not just approximately a root of $y^3 + 6y - 20$. It *is* a root! Since $y^3 + 6y - 20$ has only one real root, the complicated expression we found in Exercise 3.9 for the root as a sum of cube roots must equal 2:

$$\sqrt[3]{10 + 6\sqrt{3}} + \sqrt[3]{10 - 6\sqrt{3}} = 2.$$

This is at least a little disappointing. What if we didn't recognize that the sum of cube roots offered by Cardano's formula is simply 2? It turns out, for this example at least, that we can calculate the cube roots arising as summands in Cardano's formula explicitly as expressions involving $\sqrt{3}$ and then verify that their sum is 2. Let's do so.

Exercise 3.10. We will calculate the cube roots that appear in Exercise 3.9 by guessing and verifying.

(i) Guess that the cube root of $10 + 6\sqrt{3}$ has the form $a + b\sqrt{3}$. Cube this. Combine terms that don't involve $\sqrt{3}$ and terms that do to get two equations in a and b with integer coefficients.

(ii) There's no systematic way to solve two simultaneous equations such as these in general. Instead, turn again to guessing. Try small integers as values for a and b to find a solution.

(iii) Find a real cube root of $10 - 6\sqrt{3}$ in a similar way.

(iv) Verify that the sum of the two cube roots is 2.

Let's try another example where the same difficulty arises.

Exercise 3.11. Solve $y^3 - 2y - 4 = 0$.

(i) Use Cardano's formula to obtain

$$y = \sqrt[3]{2 + \sqrt{\frac{100}{27}}} + \sqrt[3]{2 - \sqrt{\frac{100}{27}}}.$$

(ii) Rewrite this as

$$\sqrt[3]{2 + \frac{10}{9}\sqrt{3}} + \sqrt[3]{2 - \frac{10}{9}\sqrt{3}}.$$

(iii) Taking a hint from Exercise 3.10, let's see if we can obtain a simpler expression. Find the cube root of $2 + 10\sqrt{3}/9$ by guessing that it has the form $a + b\sqrt{3}$. Cube this, set it equal to $2 + 10\sqrt{3}/9$, and obtain two equations for a and b. Try small integers for a and possibly small fractions for b to find cube roots.

(iv) Obtain a cube root of $2 - 10\sqrt{3}/9$ as well, and deduce from Cardano's formula that 2 is a solution of $y^3 - 2y - 4 = 0$.

(v) Factor $y^3 - 2y - 4$ as the product of $y - 2$ and a quadratic polynomial. Because the quadratic polynomial has no real roots, 2 is the only real root of $y^3 - 2y - 4$.

Once again, Cardano's formula has given us a solution to a cubic equation as a frightful sum of cube roots that turns out to be a simple number.

Let's consider yet another example. A surprise awaits, one of the great surprises in the history of mathematics.

Exercise 3.12. Solve $y^3 - 7y + 6 = 0$.

(i) Use Cardano's formula to obtain

$$y = \sqrt[3]{-3 + \sqrt{-\frac{100}{27}}} + \sqrt[3]{-3 - \sqrt{-\frac{100}{27}}}$$

as one root.

(ii) Rewrite this as

$$\sqrt[3]{-3 + \frac{10}{9}\sqrt{-3}} + \sqrt[3]{-3 - \frac{10}{9}\sqrt{-3}}.$$

The form of the solution in Exercise 3.12 is similar to the form of the solution in Exercise 3.11, the only essential difference being the minus sign under the square root. But what a difference! Because of the minus sign, the solution makes no sense. After all, there is no square root of -3.

Let's not worry about the meaninglessness of our solution just yet. Instead, taking a hint from Exercises 3.10 and 3.11, let's treat $\sqrt{-3}$ the way we did $\sqrt{3}$ and try to find cube roots that we might be able to add to obtain a simpler answer. Even though we don't know what $\sqrt{-3}$ means, let's assume in the exercises that when we square it, the result is the number -3.

Exercise 3.13. Let's guess that $-3 + 10\sqrt{-3}/9$ has a cube root in the form $a + b\sqrt{-3}$. We defer any concerns about the meaning of $a + b\sqrt{-3}$, choosing for now just to work with it formally.

(i) Cube $a + b\sqrt{-3}$. Combine terms that don't involve $\sqrt{-3}$ and terms that do to get two equations in a and b with integer coefficients.

(ii) Make guesses for a and b. Small positive integers for a and fractions involving thirds for b should give a solution quickly.

(iii) Find the cube root of $-3 - 10\sqrt{-3}/9$ similarly.

(iv) The expressions for the cube roots of $-3+10\sqrt{-3}/9$ and $-3-10\sqrt{-3}/9$ involve the square root of -3. Let's not worry yet. Let's proceed.

(v) Add the two cube roots: the troublesome $\sqrt{-3}$ terms cancel, leaving us with a meaningful real number, 2. Verify that 2 is a solution to $y^3 - 7y + 6 = 0$.

(vi) Conclude that Cardano's formula has led us to a correct solution of the equation $y^3 - 7y + 6 = 0$, namely, $y = 2$.

(vii) Since 2 is a solution, the polynomial $y^3 - 7y + 6$ factors as the product of $y - 2$ and a quadratic polynomial. Divide $y - 2$ into $y^3 - 7y + 6$ to find the quadratic polynomial, then determine its roots, thereby finding all three roots of $y^3 - 7y + 6$.

Cardano's formula has successfully produced the roots of $y^3 - 7y + 6$, provided we are willing to work formally with the meaningless expression $\sqrt{-3}$.

We can use Exercise 3.13 to solve the cubic equation introduced in Exercise 3.3.

Exercise 3.14. Solve

$$x^3 - 3x^2 - 4x + 12 = 0$$

by using the change of variable of Theorem 3.3 and applying the result of Exercise 3.13.

Let's conclude this section with one more illustration of what becomes possible once we open the door to $\sqrt{-3}$.

Exercise 3.15. Solve $x^3 - 1 = 0$.

(i) Observe that 1 is a solution.

(ii) Factor $x^3 - 1$ as $(x-1)(x^2 + mx + n)$ and determine m and n.

(iii) Use the quadratic formula to solve $x^2 + mx + n = 0$. Obtain in this way two additional solutions of $x^3 = 1$, the two "numbers"

$$-\frac{1}{2} \pm \frac{\sqrt{-3}}{2}.$$

(iv) For the displayed numbers to make sense, there must be a square root of -3. Let's continue not to worry about this. It will be convenient to have a special name for the hypothetical number

$$-\frac{1}{2} + \frac{\sqrt{-3}}{2}.$$

We will use the lower case Greek letter omega for this purpose and write $-1/2 + \sqrt{-3}/2$ as ω.

(v) Verify that the other hypothetical number,

$$-\frac{1}{2} - \frac{\sqrt{-3}}{2},$$

is ω^2.

(vi) Multiply ω by ω^2 to get 1. In other words, verify that $\omega^3 = 1$. Similarly, verify that $(\omega^2)^3 = 1$. Thus, ω and ω^2 are both solutions to $x^3 = 1$.

(vii) Conclude that the three cube roots of 1 are 1, ω, and ω^2.

Exercise 3.16. Let c be any non-zero real number. From Exercise 3.5, c has one real cube root. Call it a. Provided that the new number ω introduced in the preceding exercise makes sense, show that ωa and $\omega^2 a$ are also cube roots of c. Conclude that if we decide to treat ω as an allowable number, we will find that the polynomial $x^3 - c$ has three distinct roots, one real and the others of a new form.

3.3 Graphs

In this section, we will look at the shapes of cubic polynomial graphs. The results of this section are needed only in Section 3.4, where we take our first look at the discriminant of a cubic. We will return to the study of the discriminant in Section 5.3 and obtain results purely algebraically, without reference to this section. Thus, we might choose to omit both this section and Section 3.4. However, the principal results enhance our visual or geometric understanding of cubic polynomials.

To give full proofs of the results of this section, some background from the foundations of real numbers and calculus is needed. Readers with that background, including familiarity with the connection between derivatives and turning points summarized in Section 1.6, will be able to prove the results easily. For those unfamiliar with calculus, we will indicate an approach that reduces the calculus to a minimum, although some foundational theorems on the real numbers are needed.

We will restrict ourselves to reduced cubic polynomials, those of the form $x^3 + px + q$. As we saw in Section 3.1, this is in fact no restriction at all. Recall from Section 1.6 the notions of local maximum, local minimum, and turning point for the graph of a function $y = f(x)$. We know from Theorem 1.15 that the graph of $x^3 + px + q$ has at most two turning points. We know from Theorem 1.13 that the graph rises from arbitrarily low heights on the left of the y-axis to arbitrarily high heights on the right. The theorems tell us that the graph of a cubic has two possible behaviors: it will rise steadily as x increases or it will rise to a local maximum, fall to a local minimum, then rise. Which one occurs depends on the sign of p.

The possibilities are illustrated in Figure 3.1, which shows the graphs of the cubic polynomials $y^3 + px$ for $p = -4, -2, 0, 2,$ and 4. For $p = 2$ or 4, (or, more generally, p positive), the graph rises steadily, with no turning points. For $p = -2$ or -4 (or, more generally, for p negative), the graph rises, then falls, then rises. An exceptional case occurs when $p = 0$. Here, the graph never turns, but at $x = 0$, it stops rising for an instant, being essentially flat. More precisely, the x-axis is tangent to the graph at $(0, 0)$,

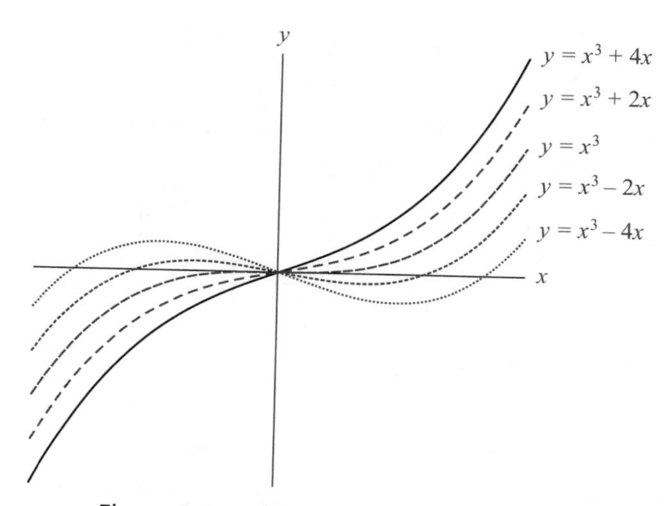

Figure 3.1. Shapes of cubic polynomial graphs

so that the slope of the graph there is 0.

Let us return to arbitrary reduced cubic polynomials.

Exercise 3.17. Let p and q be real numbers with $p \geq 0$. Let $f(x) = x^3 + px + q$.

(i) Show for any two real numbers $a < b$ that $f(a) < f(b)$.

(ii) Conclude, using Theorem 1.14, that Theorem 3.4 holds.

Theorem 3.4. *Let p and q be real numbers with $p \geq 0$. Let $f(x) = x^3 + px + q$.*

(i) The graph of $y = f(x)$ never falls as x increases. In particular, the graph of $y = f(x)$ has no turning point.

(ii) The graph crosses the x-axis exactly once, so that $f(x)$ has exactly one real root.

The graph of $y = x^3 + px + q$ crosses the y-axis at $(0, q)$. Those familiar with calculus will recognize that the slope of the graph at $(0, q)$ is p. Thus, the larger p is, the steeper the graph is at $(0, q)$ and the smaller p is, the flatter the graph is at $(0, q)$. In the extreme case that $p = 0$, the polynomial $f(x)$ is just $x^3 + q$ and the slope at $(0, q)$ is 0. Thus, the graph is flat there, with tangent line $y = q$.

When $p < 0$, the graph of $x^3 + px + q$ has two turning points. We can use elementary calculus to prove this and determine where they are:

Theorem 3.5. *Let p and q be real numbers with $p < 0$ and let $f(x) = x^3 + px + q$. Let a be the positive square root of $-p/3$.*

(i) The graph of $y = f(x)$ has two turning points, a local maximum at $x = -a$ and a local minimum at $x = a$.

(ii) As x increases, the graph rises to a turning point at

$$\left(-a, q - \frac{2ap}{3}\right),$$

falls to a turning point at

$$\left(a, q + \frac{2ap}{3}\right),$$

then rises.

Exercise 3.18. We can use elementary calculus to prove Theorem 3.5:

(i) Calculate where $f'(x)$ takes on the value 0 and deduce that $x = -a$ and $x = a$ are the only candidates for turning points.

(ii) Determine the sign of $f'(x)$ for values of x satisfying $x < -a$, $-a < x < a$, and $a < x$. Deduce that there are indeed turning points at $x = -a$ and $x = a$.

(iii) Calculate the heights $f(-a)$ and $f(a)$ of the turning points.

(iv) Deduce that the theorem holds.

We can prove much of Theorem 3.5 without calculus. Let's do so, to get an enhanced understanding of why Theorem 3.5 holds. The shape of the graph of $x^3 + px + q$ is independent of q. As q varies, the graph slides up and down (as described in Exercise 2.11), but otherwise remains unchanged.

Exercise 3.19. Let p and q be real numbers with $p < 0$ and let a be the positive square root of $-p/3$. In this exercise, we will find two turning points for the graph of $y = x^3 + px + q$. We can rewrite the cubic as $x^3 - 3a^2x + q$.

(i) Let $g(x) = x^3 - 3a^2x + 2a^3$. Because a is a root, $x - a$ divides $g(x)$. Factor $g(x)$ as $(x - a)^2(x + 2a)$. Use this factorization to verify for $x > a$ that as x increases, so does $g(x)$. Also check that $g(a) = 0$, but $g(x) > 0$ for any $x \geq 0$ besides $x = a$.

(ii) Deduce that $(a, 0)$ is a local minimum for $g(x)$ and that the graph of $g(x)$ rises to the right of $x = a$ as x increases.

(iii) Let $h(x) = x^3 - 3a^2x - 2a^3$. Because $-a$ is a root, $x + a$ divides $h(x)$. Factor $h(x)$ as $(x + a)^2(x - 2a)$. Use this factorization to verify for $x < -a$ that as x increases, so does $h(x)$. Also check that $h(-a) = 0$, but $h(x) < 0$ for any $x \leq 0$ besides $x = -a$.

(iv) Let q be a real number and let $f(x) = x^3 - 3a^2x + q$. The graph of $f(x)$ is just a vertical shift of the graphs of $g(x)$ and $h(x)$. Conclude that $f(x)$ has a local maximum at $(-a, q - 2ap/3)$ and a local minimum at $(a, q + 2ap/3)$, with the graph rising as x increases to $-a$ and as x increases from a.

Let's continue with the notation of Exercise 3.19. We might guess from the exercise that the graph of $y = x^3 + px + q$ falls as x goes from $-a$ to a. Theorem 3.5 tells us that this is the case, as we easily check using calculus, since we can verify that the derivative $3x^2 + p$ of $x^3 + px + q$ is negative for all values of x satisfying $-a < x < a$.

Without calculus, we can still show with a little more work (and an appeal to basic facts from the foundations of real numbers) that $x^3 + px + q$ decreases as x goes from $-a$ to a. For instance, we can show for each real number r between $-a$ and a that there is some open interval around r on which $x^3 + px + q$ is decreasing. This local information, together with basic results on the real numbers, shows $x^3 + px + q$ is decreasing across the entire interval $(-a, a)$. Rather than pursuing this point further, we will accept that it is true, as we already know from calculus.

3.4 A Discriminant

For a quadratic polynomial $x^2 + bx + c$, we know that there are three possibilities for the roots: two distinct real roots, one real root repeated, or no real roots. We also know that we can decide which case holds from the coefficients b and c, as we saw in Theorem 2.5. The three cases correspond to the quantity $b^2 - 4c$, the discriminant of $x^2 + bx + c$, being positive, zero, or negative. This can be proved by using elementary calculus or the quadratic formula.

For a cubic polynomial $x^3 + bx^2 + cx + d$, as for a quadratic polynomial, it is possible to determine from the coefficients the nature of the cubic's roots: whether the cubic has one simple real root, three distinct real roots, or repeated real roots. This is governed by an expression in b, c, and

d, known as the discriminant. We will discuss an algebraic approach to the discriminant in Section 5.3. An alternative approach based on an understanding of the graph of a cubic polynomial (and therefore depending on calculus) is also possible. In this section, we will describe this approach for a reduced cubic polynomial, using the results of Section 3.3.

We continue to work with a cubic polynomial x^3+px+q, where p and q are real numbers, and we will write it as $f(x)$. Our analysis, as the results of Section 3.3 would suggest, depends on whether p is positive, negative, or 0. In case $p = 0$, we analyzed the polynomial $x^3 + q$ in Exercise 3.5. It has one real root, the real cube root of $-q$, which is of multiplicity 3 if $q = 0$ and of multiplicity 1 otherwise.

Exercise 3.20. Let $f(x) = x^3 + px + q$ with $p > 0$.

(i) Recall from Theorem 3.4, or observe anew, that the graph of $y = f(x)$ is always increasing and that therefore $f(x)$ has only one real root.

(ii) Let r be the real root. From Theorem 3.2, either r has multiplicity 1 as a root of $f(x)$, in which case $f(x)$ factors as the product of $x - r$ and a quadratic polynomial that is not a product of linear polynomials, or r has multiplicity 3, in which case $f(x)$ factors as $(x - r)^3$. Show that $f(x)$ cannot factor as $(x - r)^3$ and conclude that r has multiplicity 1.

We will handle the case $p < 0$ in Exercise 3.21, using Theorem 3.5. It will be helpful to have a picture in mind of the possible behaviors of the graphs. We will use Figure 3.2 as a guide. The figure shows the graphs of the cubic polynomials $x^3 - 3x + q$ for $q = -4, -2, 0, 2,$ and 4. All five graphs

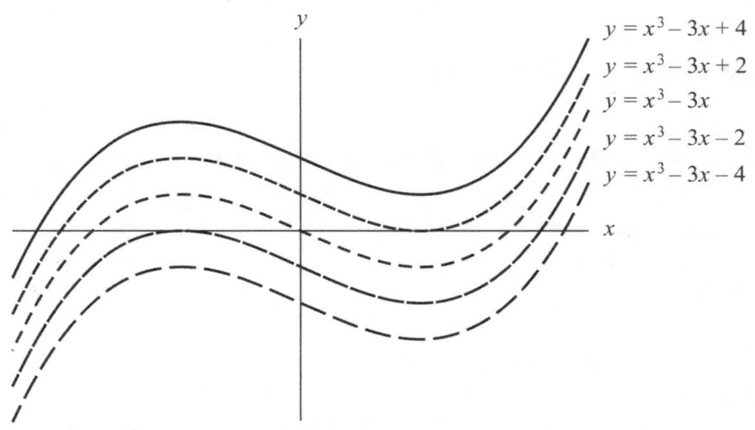

Figure 3.2. Intersection of $y = x^3 - 3x + q$ with x-axis

have the same shape, as we would expect. They are translations of each other, up or down. How far each one is shifted determines how the x-axis meets it and therefore how many roots the corresponding polynomial has.

Imagine a video showing the graph of $y = x^3 - 3x + q$ as q increases from -10 to 10. At the start of the video, the graph of $y = x^3 - 3x - 10$ is shown. As time goes by, the curve steadily rises, until at the end of the video we arrive at the graph of $y = x^3 - 3x + 10$.

As suggested in Figure 3.2 with $q = -4$, at the video's start, the curve will cross the x-axis only once. When $q = -2$, the curve will cross the x-axis off to the right, but suddenly the turning point on the left makes contact with the x-axis, producing a second root. Once q increases above -2, the turning point rises above the x-axis and is replaced by two points of intersection on the left, along with the point of intersection on the right. Thus, as the figure illustrates in the case of $q = 0$, whenever $-2 < q < 2$ the graph of $x^3 - 3x + q$ crosses the x-axis three times and there are three roots. The behavior changes when $q = 2$. If we were to watch the video as q increases from -2 to 2, we would see the middle of the three intersection points approach the rightmost intersection point, until they merge when $q = 2$. The graph of $x^3 - 3x + 2$ therefore meets the x-axis only twice and the polynomial has two roots. Finally, as q increases beyond 2, the graph no longer makes contact with the x-axis to the right, crossing only on the left, with $x^3 - 3x + q$ having only one root.

Exercise 3.21. Let $f(x) = x^3 + px + q$ with $p < 0$. Let a be the positive square root of $-p/3$.

(i) Recall from Theorem 3.5 that the graph of $y = f(x)$ has two turning points, a local maximum at $(-a, q - 2ap/3)$ and a local minimum at $(a, q + 2ap/3)$.

(ii) The real roots of $f(x)$ occur at the x-coordinates of the points where the graph of $y = f(x)$ meets the x-axis. Geometrically there are five cases, as depicted in Figure 3.2 and described in the discussion that preceded this exercise:

(a) The x-axis crosses the graph above the local maximum:

$$f(a) < f(-a) < 0.$$

(b) The x-axis crosses the graph at the local maximum:

$$f(a) < f(-a) = 0.$$

(c) The x-axis crosses the graph between the local minimum and the local maximum:

$$f(a) < 0 < f(-a).$$

(d) The x-axis crosses the graph at the local minimum:

$$0 = f(a) < f(-a)$$

(e) The x-axis crosses the graph below the local minimum:

$$0 < f(a) < f(-a).$$

(iii) For each of these five cases, describe the number of real roots of $f(x)$ and their multiplicity, referring to Theorem 3.2 for guidance on the options.

(iv) Express the last result by saying that there are three possibilities for the nature of the roots, depending on whether $f(-a)$ and $f(a)$ have opposite sign, $f(-a)$ and $f(a)$ have the same sign, or one of them is 0.

(v) The three cases can be described more compactly in terms of the sign of the product of $f(-a)$ and $f(a)$:

(a) If $f(-a)f(a) < 0$, then $f(x)$ has three distinct real roots.
(b) If $f(-a)f(a) > 0$, then $f(x)$ has one real root, of multiplicity 1.
(c) If $f(-a)f(a) = 0$, then $f(x)$ has two distinct real roots, of multiplicities 1 and 2.

(vi) Calculate the product $f(-a)f(a)$ to get

$$f(-a)f(a) = q^2 + \frac{4p^3}{27},$$

so

$$-27f(-a)f(a) = -4p^3 - 27q^2.$$

Given the cubic polynomial $x^3 + px + q$, whatever the sign of p, let's introduce a name and a notation for the quantity that appeared at the end of Exercise 3.21. Set

$$\delta = -4p^3 - 27q^2$$

and call δ the *discriminant* of $x^3 + px + q$. We will discover an algebraic interpretation of this quantity in Section 5.3.

Exercise 3.22. Let's continue the analysis of the roots of the cubic polynomial $x^3 + px + q$.

(i) Assume $p = 0$. Then $x^3 + px + q$ has one real root, which has multiplicity 1 if $q \neq 0$ and multiplicity 3 if $q = 0$. Verify that $\delta < 0$ if $q \neq 0$ and $\delta = 0$ if $q = 0$.

(ii) Assume $p > 0$. From Exercise 3.20, $x^3 + px + q$ has one simple real root. Verify that $\delta < 0$.

(iii) Assume $p < 0$. Deduce from Exercise 3.21 that when the graph of $y = x^3 + px + q$ crosses the x-axis three times, so that $x^3 + px + q$ has three distinct real roots, $\delta > 0$. Deduce that when the graph crosses the x-axis only once, so that $x^3 + px + q$ has one real root, $\delta < 0$. Deduce that when the graph crosses the x-axis once and is tangent to the x-axis once, so that there is a repeated real root, $\delta = 0$.

We have considered all possible cases. Reviewing them, we see that δ is positive precisely when there are three distinct real roots, negative precisely when there is only one simple real root, and 0 when there is a repeated real root. Thus, we have proved the following theorem:

Theorem 3.6. *Let $f(x) = x^3 + px + q$ for real numbers p and q. Let $\delta = -4p^3 - 27q^2$, the discriminant of $f(x)$.*

(i) If $\delta > 0$, then $f(x)$ has three distinct real roots.

(ii) If $\delta = 0$, then $f(x)$ has a repeated real root.

(iii) If $\delta < 0$, then $f(x)$ has one simple real root.

Exercise 3.23. Using Theorem 3.6, describe the nature of the roots of each of these polynomials:

(i) $x^3 - 3x + 2$.

(ii) $x^3 + 6x - 20$.

(iii) $x^3 - 2x - 4$.

(iv) $x^3 - 7x - 6$.

Cardano's formula for a root of the reduced cubic polynomial $x^3 + px + q$ includes the square root of the quantity $p^3/27 + q^2/4$, which we denoted by R. The discriminant δ and R are constant multiples of each other:

$$\delta = -108R.$$

If R is negative, Cardano's formula contains square roots of negative numbers. Since R is negative when δ is positive, and δ is positive precisely when $f(x)$ has three distinct real roots, we see that the troublesome situation of having to deal with square roots of negative numbers in using Cardano's formula arises precisely when $f(x)$ has three distinct real roots. We will discuss this further in Section 5.4.

3.5 History

We closed our account of quadratic equations in Section 2.5 with Luca Pacioli, who summarized the mathematical knowledge of the time in 1494 [50]. Regarding cubic and quartic equations, Pacioli wrote [66, p. 47] that "it has not been possible until now to form general rules." This was the setting at the dawn of the sixteenth century.

By the century's end, general rules would be in place, thanks to the work of Cardano and other Italian mathematicians. The story of these discoveries is a wonderful one. Given the excellent accounts in [67] and [66], as well as the 1953 biography *Cardano: The Gambling Scholar* [49] by the mathematician Øystein Ore, only a brief account will be given here. (Ore's biography is unfortunately no longer in print. For more on Cardano, one can also turn to Anthony Grafton's *Cardano's cosmos: the worlds and works of a Renaissance astrologer* [33] as well as Cardano's autobiography, *The Book of My Life* [13], available in a 2002 edition that contains both Jean Stoner's 1929 English translation and an introduction by Grafton.)

An important point to keep in mind in preparation for the story is that the academic culture at the time was not anything like that to which we are accustomed. If someone were now to solve a centuries-long problem, he or she would immediately announce it, lecture on it, and publish it, with the written account being made available on the internet long before it actually appears in print. Other mathematicians would study the solution, make sure there are no errors, and pay close attention to the methods used for the solution, in anticipation that the methods and the new ideas they contain may be applicable to other problems. In contrast, in sixteenth century Italy, scholars would keep techniques of solution to themselves, perhaps employing them to succeed at public competitions.

Scipione del Ferro was the first person to obtain a formula for solutions to cubic equations. Del Ferro lived from 1465 until 1526, serving as a professor at the University of Bologna for the final thirty years of his life. He found a way to solve cubic equations of the form

$$x^3 + px = q.$$

Del Ferro and his contemporaries worked with positive numbers only, both as coefficients and as solutions, so p and q are understood to be positive and we seek positive values of x. An example would be $x^3 + 6x = 20$, the cubic equation of Exercise 3.9.

We saw in Section 2.5 that seven centuries earlier, al-Khwarizmi had classified quadratic equations into three forms (excluding cases in which one of the coefficients is 0). Similarly, because of the restriction to positive coefficients, there are three forms for cubic equations with no degree 2 term, the form $x^3 + px = q$ that del Ferro studied as well as

$$x^3 = px + q$$

and

$$x^3 + q = px.$$

Were 0 available, we might momentarily be tempted to add a fourth case,

$$x^3 + px + q = 0.$$

But since p and q are positive, it cannot have a positive solution. Hence, we are missing nothing by omitting it.

Del Ferro did not publish his solution, but he did communicate it to his son-in-law, Annibale della Nave, and his colleague Antonio Maria Fior. In 1535, Fior challenged the Venetian mathematician Niccolò Fontana (1499–1557) to a contest in which each would pose thirty problems to the other, the loser paying for a banquet for thirty. Fontana, better known by his nickname Tartaglia, or the stutterer, prepared problems of varying types. In contrast, all of Fior's problems were cubic equations of the type $x^3 + px = q$, with del Ferro's solution as his secret weapon.

Just before time expired, on the night that ran from February 12 to 13, Tartaglia found a way to solve the equation $x^3 + px = q$ on his own! So much for Fior's secret weapon. Tartaglia solved all thirty of Fior's problems, whereas Fior could solve only some of Tartaglia's. Victory was sufficient satisfaction for Tartaglia, who chose to forgo the banquet. Of course, he had also won eternal fame as independent co-discoverer, with del Ferro, of the cubic's solution.

The story now shifts to another participant, the one for whom the solution to the cubic equation is named, Girolamo Cardano. Cardano was a prominent scholar in many fields, famous as a doctor, astrologer, philosopher, and mathematician. He lived from 1501 to 1576 and spent most of his life in the city of his birth, Milan. In 1539, having heard of Tartaglia's solution to the cubic, he asked Tartaglia through an intermediary what the solution was, but Tartaglia chose not to tell. Cardano then invited Tartaglia to come to Milan as his guest, enticing him with the opportunity to meet Alfonso d'Avalos, the military commander of Milan, to whom Tartaglia would be able to show some of his military inventions.

Like Cardano, Tartaglia was a man of many talents, including expertise in ballistics and military engineering. In 1546, he would publish some of his military work in *Quesiti et Inventioni Diverse*, or *New Problems and Inventions* [63]. (Its drawings of cannons, cannonball paths, and fortifications are reason enough to seek out a copy of the book.) Naturally, Tartaglia accepted Cardano's invitation. During his stay, Tartaglia told Cardano how he solved the equation $x^3 + px = q$, with Cardano swearing an oath on March 29, 1539 never to publish it.

We should note at this point that the principal source for many details of this story, including Cardano's oath, is Tartaglia himself, as he would devote the final pages of *Quesiti et Inventioni Diverse* to an account of his dealings with Cardano.

After Tartaglia left, Cardano saw how to use Tartaglia's ideas in order to obtain solutions to the cubic equations of the forms $x^3 = px + q$ and $x^3 + q = px$. Difficulties arise in these two cases because for certain values of p and q, the solutions may contain expressions with square roots of negative numbers, as we saw in Exercise 3.12. (In contrast, this cannot occur in the expression for a solution of the cubic equation $x^3 + px = q$.) Nonetheless, in principle Cardano had found a solution to any cubic equation without a degree 2 term, and as we know, it is then an elementary matter to solve all cubic equations.

The fourth participant in our story is Lodovico Ferrari, who lived from 1522 to 1565. Ferrari came from Bologna to Milan at the age of 14 to work as a servant in Cardano's household. Cardano quickly realized Ferrari's talent, and Ferrari moved from servant to student to collaborator. Ferrari learned from Cardano the method of solving cubic equations, then made his own great contribution: the solution of quartic equations. What Ferrari discovered is that one can reduce the problem of solving a quartic equation to that of solving an auxiliary cubic equation, one whose coefficients are expressed in terms of the coefficients of the original quartic. We will study Ferrari's discovery in Section 6.2.

Ferrari's result was an advance of the greatest importance, but it posed a difficulty for Cardano. He had given his oath to Tartaglia that he would not publish Tartaglia's solution to cubic equations of a special form. Yet, Cardano had extended this to other cubic equations and his disciple Ferrari had shown how to use it to solve quartic equations. This was too important to keep secret. Moreover, he may have had a way out, since the formula was initially due to del Ferro, not Tartaglia.

Cardano decided to publish the solution, which he did in the book *Ars Magna*, written in Latin and published in 1545. (It is available in an English translation by T. Richard Witmer, with the title *The Great Art; or, The Rules of Algebra* [12].) He states clearly at the beginning of *Ars Magna* [12, pp. 8–9] that the solution to the cubic equation in the special form $x^3 + px = q$ was discovered by del Ferro and re-discovered by Tartaglia:

> In our own days Scoipione del Ferro of Bologna has solved the case of the cube and first power equal to a constant, a very elegant and admirable accomplishment. Since this art surpasses all human subtlety and the perspicuity of mortal talent and is a truly celestial gift and a very clear test of the capacity of men's minds, whoever applies himself to it will believe that there is nothing he cannot understand. In emulation of him, my friend Niccolò Tartaglia of Brescia, wanting not to be outdone, solved the same case when he got into a contest with his [Scipioine's] pupil, Antonio Maria Fior, and, moved by my many entreaties, gave it to me. For I had been deceived by the words of Luca Paccioli, who denied that any more general rule could be discovered than his own. Notwithstanding the many things which I had already discovered, as is well known, I had despaired and had not attempted to look any further. Then, however, receiving Tartaglia's solution and seeking for the proof of it, I came to understand that there were a great many other things that could also be had. Pursuing this thought and with increased confidence, I discovered these others, partly by myself and partly through Lodovico Ferrari, formerly my pupil. Hereinafter those things which have been discovered by others have their names attached to them; those to which no name is attached are mine. The demonstrations, except for the three by [al-Khwarizmi] and the two by Lodovico, are all mine.

Cardano turns to cubic equations in Chapter 11, "On the Cube and First Power Equal to the Number," which is devoted to equations of the form $x^3 + px = q$. At the start of the chapter, he once again credits del Ferro and Tartaglia [12, p. 96]:

> Scipio Ferro of Bologna well-nigh thirty years ago discovered this rule and handed it on to Antonio Maria Fior of Venice, whose contest with Niccolò Tartaglia of Brescia gave Niccolò occasion to discover it. He [Tartaglia] gave it to me in response to my entreaties though withholding the demonstration. Armed with this assistance, I sought out its

> demonstration in [various] forms. This was very difficult. My version of it follows.

By modern standards, Cardano had given proper credit for the results on which he and Ferrari built, and had satisfied all scholarly expectations. But this was a different era, and Tartaglia was furious. As already noted, he told his side of the story a year later in *Quesiti et Inventioni Diverse* [63], making the story of the oath—and its text—public.

Cardano did not respond to Tartaglia's accusations, but Ferrari, on Cardano's behalf, issued a public challenge. The dispute was argued in Milan on August 10, 1548, with Tartaglia leaving before it was settled and evidently being named the loser. (See Ore's discussion [49, pp. 99–105] for an attempt to sort out the details from the available records.)

These discoveries make for a richly entertaining story, worthy of the greatness of the discoveries themselves. As Varadarajan notes in concluding his account of the dispute [67, p. 62], "Tartaglia, through his penchant for secrecy, represents the middle ages, while Cardano, with his generous views about authorship and sharing of knowledge, represents the modern view. Mathematics was very fortunate that a person with the great vision and world-view like Cardano was able to get his hands on the original discovery and was then able to build a wonderful structure on top of it that led to the beginning of Algebra as we know today." Ore offers the following conclusion [49, p. 106]:

> Cardano and Ferrari represent by far the greater mathematical penetration and wealth of novel ideas. Tartaglia was also, doubtless, an excellent mathematician, but his great tragedy was the head-on collision with the only two opponents in the world who could be ranked above him. Without them, he would most likely have been reckoned as the foremost mathematician in the middle of the sixteenth century.

As for Cardano's actual account of his formula, let us see what he says in Chapter 11 of *Ars Magna* in his treatment of equations of the form $x^3 + px = q$ [12, pp. 98–99]. Some of his terminology has been modernized in the English translation. The terms *binomium* and *apotome*, which the translation retains, refer to pairs of numbers such as $\sqrt{5} + 3$ and $\sqrt{5} - 3$.

> Cube one-third the coefficient of x; add to it the square of one-half the constant of the equation; and take the square root of the whole. You will duplicate this, and to one of the two you add one-half the number you have already squared and from the other you subtract one-half the same. You then have a *binomium* and its *apotome*. Then, subtracting

the cube root of the *apotome* from the cube root of the *binomium*, the remainder [or] that which is left is the value of x.

For example,

$$x^3 + 6x = 20.$$

Cube 2, one-third of 6, making 8; square 10, one-half of the constant; 100 results. Add 100 and 8, making 108, the square root of which is $\sqrt{108}$. This you will duplicate: to one add 10, one-half the constant, and from the other subtract the same. Thus you will obtain the *binomium* $\sqrt{108} + 10$ and its *apotome* $\sqrt{108} - 10$. Take the cube roots of these. Subtract [the cube root of the] *apotome* from that of the *binomium* and you will have the value of x:

$$\sqrt[3]{\sqrt{108} + 10} - \sqrt[3]{\sqrt{108} - 10}.$$

One should compare this description with Exercise 3.9.

Cardano continues in Chapter 11 with two more examples [12, pp. 99–100], the equations $x^3 + 3x = 10$ and $x^3 + 6x = 2$. He then returns to the first example and notes that the complicated difference of cube roots can be calculated to be 2.

We will return to the cubic equation's history in Section 5.8. We conclude for now with the opening words of Grafton's introduction to Cardano's autobiography [13, p. ix]:

> Cardano dazzled readers across sixteenth-century Europe. His original and influential books dealt with medicine, astrology, natural philosophy, mathematics, and morals—to say nothing of devices for raising sunken ships and stopping chimneys from smoking. They won the attention of popes and inquisitors, Catholics and Protestants, theologians and playwrights. And nothing did more to enhance their appeal than the polished stories about Cardano's past that glittered like enticing gems in his most technical treatises, or the longer, dramatic retellings of his whole life story that he offered the public at suitable intervals.

4

Complex Numbers

Our experience with Cardano's formula has taught us that to solve cubic equations, we need to work with numbers such as $\sqrt{-3}$, that is, square roots of negative numbers. Cubic equations were where they were first encountered. They are now called complex numbers, and have been found to have many important uses in mathematics and science. In this chapter, we will introduce them and see how to calculate their nth powers and roots, which is needed in our study of polynomial equations.

4.1 Complex Numbers

In this section we will develop the arithmetic of complex numbers. We begin by introducing a new number, $\sqrt{-1}$, to serve as a square root of -1. Once $\sqrt{-1}$ is admitted to the roll of allowable numbers, we are led to more new numbers, such as $3\sqrt{-1}$, $\sqrt{2}+\sqrt{-1}$, and $4-\pi\sqrt{-1}$, when we try to perform arithmetic operations that combine $\sqrt{-1}$ and our old numbers. In general, given real numbers a and b, we must admit to the roll of numbers the new number $a+b\sqrt{-1}$. Numbers of this type are called *complex numbers*. Real numbers are themselves complex numbers: the real number a can be thought of as the complex number $a+0\sqrt{-1}$.

Given a complex number $a+b\sqrt{-1}$, the real number a is called its *real* part and the real number b is called its *imaginary* part. For example, the real part of $2+3\sqrt{-1}$ is 2 and the imaginary part is 3. Real numbers are complex numbers with imaginary part 0. Also of note are the complex numbers with real part 0, numbers of the form $b\sqrt{-1}$, called *pure imaginary* numbers.

Suppose we are given two complex numbers, $a+b\sqrt{-1}$ and $c+d\sqrt{-1}$. How do we add and multiply them? Let's do what comes naturally and see what happens. This will lead us to definitions of sum and product.

For addition, if we rearrange and combine terms, we obtain

$$\begin{aligned}(a+b\sqrt{-1})+(c+d\sqrt{-1}) &= a+c+b\sqrt{-1}+d\sqrt{-1}\\ &= (a+c)+(b+d)\sqrt{-1}.\end{aligned}$$

This would appear to be a sensible definition of addition, and it is what we adopt. For example,

$$(2+3\sqrt{-1})+(8+2\sqrt{-1}) = 10+5\sqrt{-1}.$$

How about multiplication? Let's use the distributive law and multiply the same two numbers:

$$(2+3\sqrt{-1})\cdot(8+2\sqrt{-1}) = 2\cdot 8+2\cdot 2\sqrt{-1}+3\sqrt{-1}\cdot 8+3\sqrt{-1}\cdot 2\sqrt{-1}.$$

It is natural to expect that $\sqrt{-1}$ should commute with real numbers so that $\sqrt{-1}\cdot 8 = 8\cdot\sqrt{-1}$. Assuming this and rearranging terms, we obtain

$$(2+3\sqrt{-1})\cdot(8+2\sqrt{-1}) = 2\cdot 8+2\cdot 2\sqrt{-1}+3\cdot 8\sqrt{-1}+3\cdot 2(\sqrt{-1})^2.$$

If $\sqrt{-1}$ is to make sense as a number, then it must have the property that its square is -1:

$$(\sqrt{-1})^2 = -1.$$

If we assume this, then we find that

$$(2+3\sqrt{-1})\cdot(8+2\sqrt{-1}) = 16+4\sqrt{-1}+24\sqrt{-1}+6(-1) = 10+28\sqrt{-1}.$$

More generally, suppose we wish to multiply $a+b\sqrt{-1}$ and $c+d\sqrt{-1}$. Proceeding in the same way, assuming $\sqrt{-1}\cdot c = c\cdot\sqrt{-1}$ and $(\sqrt{-1})^2 = -1$, we will obtain

$$(a+b\sqrt{-1})\cdot(c+d\sqrt{-1}) = (ac-bd)+(ad+bc)\sqrt{-1}.$$

Let us adopt this as our definition of multiplication. The product of two complex numbers is again a complex number, as would need to be the case if the definition is to be of any use.

Let's summarize. Given complex numbers $a+b\sqrt{-1}$ and $c+d\sqrt{-1}$, we have adopted the rules that their sum and product are given by

$$(a+b\sqrt{-1})+(c+d\sqrt{-1}) = (a+c)+(b+d)\sqrt{-1}$$

and

$$(a+b\sqrt{-1})\cdot(c+d\sqrt{-1}) = (ac-bd)+(ad+bc)\sqrt{-1}.$$

Exercise 4.1. Our definition of multiplication for two complex numbers was motivated by the assumptions that $\sqrt{-1}$ commutes with real numbers and that $(\sqrt{-1})^2 = -1$. Now that we've defined multiplication, we should verify that these properties actually hold.

(i) Given a real number a, use the definition of multiplication to prove that

$$a \cdot \sqrt{-1} = \sqrt{-1} \cdot a.$$

(Hint: The real number a is the complex number $a + 0 \cdot \sqrt{-1}$, while $\sqrt{-1}$ is the complex number $0 + 1 \cdot \sqrt{-1}$.)

(ii) Use the definition of multiplication to prove that

$$(\sqrt{-1})^2 = -1.$$

Often the symbol i is used in place of $\sqrt{-1}$ to denote the square root of -1. There's no need to do this, but it does make our expressions easier to write. In this notation, a complex number has the form $a + bi$, for real numbers a and b, and i satisfies $i^2 = -1$. The sum of two complex numbers $a + bi$ and $c + di$ is

$$(a + bi) + (c + di) = (a + c) + (b + d)i$$

and their product is

$$(a + bi)(c + di) = (ac - bd) + (ad + bc)i.$$

Exercise 4.2. Add and multiply the pairs of complex numbers.

(i) 3 and $17 - 15i$.

(ii) $3i$ and $17 - 15i$.

(iii) i and $-i$.

(iv) $4 + i$ and $3 + 7i$.

(v) $-4 + 5i$ and $-3 + 6i$.

(vi) $2 + 3i$ and $2 - i$.

(vii) $5 + 2i$ and $5 - 2i$.

A positive real number r has a real square root $\sqrt{r}$. When we square a pure imaginary number $\sqrt{r} \cdot i$ according to our multiplication rule, we obtain

$$(\sqrt{r} \cdot i)^2 = -r.$$

Thus $\sqrt{r} \cdot i$ is a square root of $-r$. By creating the new number i to serve as a square root of -1, we have obtained square roots for all negative real numbers.

A notion that we will need is that of complex conjugate. The *complex conjugate*, or simply the *conjugate*, of a complex number $a + bi$ is the complex number $a - bi$. The conjugate of a complex number r is denoted by putting a bar over r to get $\overline{r}$. In this notation, we would write for example that

$$\overline{2 + 3i} = 2 - 3i$$

and

$$\overline{-12i} = 12i.$$

The complex conjugate of a real number a is a itself and the complex conjugate of a pure imaginary number bi is its opposite $-bi$.

A *multiplicative inverse* of a number r is a number s satisfying $rs = 1$. For instance, the multiplicative inverse of -1 is -1 and the multiplicative inverse of 2 is $1/2$. What is the multiplicative inverse of π? It is $1/\pi$, of course. But this isn't an answer. It's a notation. The fact that there is a real number s for which $\pi \cdot s = 1$ is by no means obvious. It requires a proof, one that depends on foundational results about the construction of real numbers. (See, for instance, [38, pp. 71–73] for a discussion.)

Let us accept the truth of the statement that every non-zero real number r has a multiplicative inverse, which we will write as $1/r$. Thanks to conjugation, we can deduce from this that non-zero complex numbers have multiplicative inverses.

Exercise 4.3. Let a and b be real numbers.

(i) Check that $(a + bi)(a - bi) = a^2 + b^2$.

(ii) Assume that $a + bi \neq 0$. Describe (in terms of a and b) the real and imaginary parts of a complex number that is a multiplicative inverse of $a + bi$.

Let's derive the basic facts on conjugation and the arithmetic operations for complex numbers.

Exercise 4.4. Let r and s be complex numbers.

(i) Show that

$$\overline{r} + \overline{s} = \overline{r + s}$$

and

$$\overline{r} \cdot \overline{s} = \overline{rs}.$$

(ii) Use this to show that

$$\overline{r}^2 = \overline{r^2}$$

and

$$\overline{r}^3 = \overline{r^3}.$$

(iii) We can use mathematical induction to show more generally that

$$\overline{r}^n = \overline{r^n}$$

for any positive integer n.

Let's restate one part of Exercise 4.4 as a theorem, for later reference.

Theorem 4.1. *Let r and s be complex numbers.*

(i) If $r^2 = s$, then $\overline{r}^2 = \overline{s}$.

(ii) If $r^3 = s$, then $\overline{r}^3 = \overline{s}$.

One consequence of Exercise 4.4 is Theorem 4.2. We won't use it until Exercise 7.1, yet its proof is worth working out now as an exercise on conjugation facts. For quadratic, cubic, and quartic polynomials, we will obtain Theorem 4.2 independently by direct determination of the roots.

Theorem 4.2. *Let $f(x)$ be a positive-degree polynomial with real coefficients and let r be a root of $f(x)$ that is a non-real complex number. Then $\overline{r}$ is also a root of $f(x)$.*

Exercise 4.5. Prove Theorem 4.2. (Hint: Apply Exercise 4.4 to calculate $f(\overline{r})$.)

4.2 Quadratic Polynomials and the Discriminant

We can use complex numbers to obtain a better understanding of the roots of quadratic polynomials.

Exercise 4.6. Consider the quadratic polynomial $x^2 + bx + c$, with real numbers b and c as coefficients. Assume that $b^2 - 4c < 0$. In this case $x^2 + bx + c$ has no real roots.

(i) The quadratic formula gives as roots of $x^2 + bx + c$ the two numbers

$$-\frac{b}{2} \pm \frac{\sqrt{b^2 - 4c}}{2}.$$

Until now, we have viewed these as non-existent, since we had no numbers available to serve as square roots of the negative real number $b^2 - 4c$. With the introduction of complex numbers, we now have square roots.

(ii) Rewrite the two roots of $x^2 + bx + c$ given in the previous part as complex numbers, using i.

(iii) Substitute the resulting complex numbers into $x^2 + bx + c$ and verify that they are roots of $x^2 + bx + c$.

(iv) Draw the conclusion that if complex numbers are allowed, the quadratic formula makes sense even when $b^2 - 4c < 0$.

(v) The two roots we have obtained in this way are complex conjugates of each other. Write the roots as r and $\overline{r}$. Verify that

$$x^2 + bx + c = (x - r)(x - \overline{r})$$

by multiplying out the right side of the equation.

(vi) Deduce that Theorem 4.3, a refinement of Theorem 2.2, holds.

Theorem 4.3. *Let $f(x) = x^2 + bx + c$ for real numbers b and c. Exactly one of three possibilities occurs:*

(i) $x^2 + bx + c$ has two distinct real roots a_1 and a_2, and factors as

$$(x - a_1)(x - a_2).$$

(ii) $x^2 + bx + c$ has only one real root a, and factors as $(x - a)^2$.

(iii) $x^2 + bx + c$ has two distinct complex roots r and $\overline{r}$, and factors as $(x - r)(x - \overline{r})$.

Moreover, the first possibility occurs if $b^2 - 4c > 0$, the second possibility occurs if $b^2 - 4c = 0$, and the third possibility occurs if $b^2 - 4c < 0$.

We saw in Theorem 2.5 that the nature of the roots of a quadratic polynomial $x^2 + bx + c$ is determined by the sign of $b^2 - 4c$, which we denoted δ. In the next exercise, we will see that the nature is determined by the sign of the square of the difference of the polynomial's roots.

Exercise 4.7. Suppose b and c are real numbers. Write r_1 and r_2 for the roots of $x^2 + bx + c$ and Δ for $(r_1 - r_2)^2$.

(i) Conclude from Theorem 4.3 that there are three mutually exclusive possibilities for r_1 and r_2: they are real and distinct, or real and coincident, or distinct complex conjugates of each other.

(ii) If $r_1 = r_2$, then $\Delta = 0$.

(iii) If r_1 and r_2 are real and distinct, then $\Delta > 0$.

(iv) The only remaining possibility is that r_1 and r_2 are not real but are complex numbers, each the conjugate of the other. Suppose $r_1 = m + ni$ and $r_2 = m - ni$, with m and n real numbers and with $n \neq 0$. (If $n = 0$, the roots are real.) Calculate Δ and show that it is a negative real number.

(v) We wish to show that the converses of the results in (ii)-(iv) hold as well; that is:

  (a) If $\Delta > 0$, then r_1 and r_2 are real and distinct;

  (b) if $\Delta = 0$, then r_1 is a real number and $r_1 = r_2$;

  (c) if $\Delta < 0$, then r_1 and r_2 are non-real, complex numbers, each the conjugate of the other.

(vi) Assume that $\Delta > 0$. We are to prove that r_1 and r_2 are real and distinct. Suppose this is not the case and deduce from the first part of the exercise that either r_1 and r_2 coincide and are real or they are non-real complex conjugates. Use other parts of the proof to obtain the contradiction that $\Delta \leq 0$. Conclude, as desired, that r_1 and r_2 are real and distinct.

(vii) Make similar arguments to prove the other two converses.

(viii) Conclude that the sign of Δ determines the nature of the roots of $x^2 + bx + c$.

Both $b^2 - 4c$, which we write as δ, and $(r_1 - r_2)^2$, which we write as Δ, determine the nature of the roots of $x^2 + bx + c$. Let's compare the two quantities.

Exercise 4.8. With the notation of Exercise 4.7, recall that $x^2 + bx + c$ factors as $(x - r_1)(x - r_2)$. Use this to express b and c in terms of r_1 and r_2 and deduce that $(r_1 - r_2)^2 = b^2 - 4c$, so that Δ and δ coincide.

We see that the discriminant of $x^2 + bx + c$ has two descriptions. It is the square of the difference of the roots and it is the quantity $b^2 - 4c$. We are free to take either of these as the definition of the discriminant. However, the more fundamental quantity is $(r_1 - r_2)^2$. One reason is that it can be generalized for higher-degree polynomials. Hence, it is the one we adopt as the defining expression for a quadratic polynomial's discriminant.

The fact that $(r_1 - r_2)^2$ equals $b^2 - 4c$ can then be interpreted as the statement that we can calculate the discriminant of $x^2 + bx + c$, defined as the square of the difference of the roots, in terms of the coefficients b and c, without knowledge of the roots. We record this in the next theorem.

Theorem 4.4. *Let b and c be real numbers and let r_1 and r_2 be the roots (real or complex) of the quadratic polynomial $x^2 + bx + c$, so that $x^2 + bx + c$ factors as $(x - r_1)(x - r_2)$. Let Δ be the discriminant of $x^2 + bx + c$, which by definition is $(r_1 - r_2)^2$.*

(i) Δ determines the nature of the roots of $x^2 + bx + c$: If $\Delta > 0$, then the roots are real and distinct; if $\Delta = 0$, then there is one root, real of multiplicity 2; if $\Delta < 0$, then the roots are a pair of non-real complex conjugates.

(ii) Δ can be calculated in terms of the coefficients:

$$\Delta = b^2 - 4c.$$

Complex numbers have allowed us to obtain the refinement Theorem 4.3 of Theorem 2.2. Similarly, we can refine Theorem 3.2 on the roots of cubic polynomials.

Theorem 4.5. *Let $f(x)$ be a monic, cubic polynomial. Exactly one of* (i)–(iv) *occurs.*

(i) $f(x)$ has one real root a, of multiplicity 3, and factors as

$$(x - a)^3.$$

(ii) $f(x)$ has two distinct real roots a_1 and a_2, of multiplicities 1 and 2, and factors as

$$(x - a_1)(x - a_2)^2.$$

(iii) $f(x)$ has three distinct simple real roots a_1, a_2, and a_3, and factors as

$$(x - a_1)(x - a_2)(x - a_3).$$

(iv) $f(x)$ has one real root a and two distinct non-real, complex roots r and $\overline{r}$, and factors as

$$(x-a)(x-r)(x-\overline{r}).$$

Exercise 4.9. Prove Theorem 4.5.

Exercise 4.10. Suppose $x^3 + px + q$ is a cubic polynomial. We saw in Theorem 3.1 that it must have a real root. Suppose you have found such a root, r, by using Cardano's formula or other methods. Explain how you can find the other two roots, whether real or complex.

4.3 Square and Cube Roots

At the end of Section 2.1, we saw that the quadratic formula reduces the problem of solving a quadratic equation to that of calculating square roots of real numbers. At the beginning of Section 3.2, we hoped that solving cubic equations could be reduced, similarly, to the problem of calculating square and cube roots of real numbers. Our initial experience with Cardano's formula suggested that we might need to find cube roots of non-real complex numbers as well. Let's take a look at the problem of calculating square and cube roots of complex numbers.

We begin with square roots. How do we find square roots of a complex number $a + bi$? There are two issues here. Does $a + bi$ have another complex number as its square root? If so, can we calculate it? Regarding the second question, we already have noted that algebraic techniques alone do not allow us to compute square roots of positive real numbers. We must use techniques of approximation. We cannot expect the calculation of square roots of complex numbers to be any easier. We should be content if we can reduce the problem of calculating square roots of complex numbers to the problem of calculating square roots of positive real numbers.

Fix real numbers a and b; they should be regarded as known constants, not as variables. Suppose we wish to find square roots of $a + bi$. If $b = 0$, then finding the square roots of $a + bi$ reduces to finding the square roots of the real number a. This is trivial if $a = 0$. If $a \neq 0$, it reduces to the problem of finding the square root of a if a is positive or the square root of $-a$ if a is negative (in which case the square roots of a are i times the square roots of $-a$). Thus, we can assume that $b \neq 0$. Let's deal with this case in an exercise.

Exercise 4.11. Let a and b be real numbers with $b \neq 0$. We wish to find a complex number $r + si$ such that $(r + si)^2 = a + bi$. We can introduce variables x and y for the unknown real numbers r and s, so that finding a square root amounts to solving the equation

$$(x + yi)^2 = a + bi$$

for real values of the unknowns x and y. We view a and b as known, fixed constants.

(i) Expand the equation $(x + yi)^2 = a + bi$ and obtain two simultaneous equations for x and y with real coefficients.

(ii) Using one of the equations and the fact that $b \neq 0$, show that if these equations have solutions, x and y will be non-zero.

(iii) Using this, divide both sides of one equation by x to get an equation expressing y in terms of x and substitute this into the other equation to get a single equation in x.

(iv) To find a square root of $a + bi$, we need to solve a single equation in x involving only real numbers. Clear denominators in this equation and obtain a degree 4 or quartic equation in x.

(v) It can be regarded as a degree 2 or quadratic equation in x^2. Use the quadratic formula to obtain two values for x^2 in terms of a and b.

(vi) We are looking for a real value of x that solves the equation. If one of our expressions for x^2 is a positive real number, then its two square roots are our desired values for x. Verify that one of the expressions for x^2 is positive.

(vii) Take square roots to obtain two values of x. Using an equation you obtained earlier in the problem, obtain the two corresponding values for y.

(viii) Write the formulas for the two square roots of $a + bi$. Notice that each is -1 times the other.

We have just shown that any non-zero complex number has two complex numbers as its square roots, each the opposite of the other. We will record part of this.

Theorem 4.6. *Given a complex number s, there exists a complex number r satisfying $r^2 = s$; in other words, every complex number has a complex square root.*

The existence of square roots allows us to extend the quadratic formula to quadratic polynomials with complex coefficients. We won't need the result for our analysis of cubic and quartic polynomials, but we will use it in the final section, Section 7.5. In anticipation of that, let's work out the details here.

Exercise 4.12. Let b and c be complex numbers.

(i) Rework Exercise 2.7 to show that the roots of $x^2 + bx + c$ are

$$x = -\frac{b}{2} \pm \frac{\sqrt{b^2 - 4c}}{2}.$$

(ii) Use Theorem 4.6 to deduce that $x^2 + bx + c$ has a root in the complex numbers, and so factors as a product of two linear polynomials.

(iii) Conclude that Theorem 4.7 holds.

Theorem 4.7. *Let b and c be complex numbers. There exist complex numbers r_1 and r_2 (possibly coincident) such that*

$$x^2 + bx + c = (x - r_1)(x - r_2).$$

In Exercise 4.11, we have proved that complex square roots exist and shown how to determine them explicitly in terms of square roots of positive real numbers. We wish to be able to compute cube roots of complex numbers as well, since this is essential in using Cardano's formula. Our experience in Exercise 4.11 suggests that we should be able to calculate cube roots of complex numbers in terms of cube roots of real numbers. Let's mimic what we did for square roots and see how far we can get.

Exercise 4.13. Fix real numbers a and b; they should be regarded as constants, not as variables. Assume $b \neq 0$. We wish to determine the cube roots of the non-real complex number $a + bi$; that is, we wish to find a complex number $m + ni$ such that $(m + ni)^3 = a + bi$. We can introduce variables x and y for the unknown real numbers m and n, so that finding a cube root amounts to solving the equation

$$(x + yi)^3 = a + bi$$

for real values of x and y.

Expand the equation $(x + yi)^3 = a + bi$ and obtain two simultaneous equations for x and y with real coefficients. Conclude that the problem of finding a complex cube root of $a+bi$ is equivalent to the problem of finding real solutions to the two equations.

In Exercise 3.13, we solved the equations that arise in Exercise 4.13 by trial and error for certain choices of a and b. Now we want a general solution, but there is no obvious way to proceed. Let's experiment.

Exercise 4.14. Begin with the two equations for x and y obtained in Exercise 4.13.

(i) Using one of the equations and the assumption that $b \neq 0$, show that if these equations have solutions, then y will be non-zero.

(ii) Introduce a new variable s and set $x = sy$, so that $s = x/y$. This makes sense, since we know that y can't be 0. Substitute sy for x in the two equations for x and y to get two equations in s and y. Write them so that each is an expression in s times y^3. Using the assumption that $b \neq 0$, show that $3s^2 - 1 \neq 0$.

(iii) Solve one equation for y^3, substitute in the other, and obtain a cubic equation in s with coefficients involving a, b, and 3.

We have reduced the problem of finding a cube root of $a + bi$ to the problem of solving a cubic equation in s with real coefficients. If we can solve the cubic equation for s, then we can determine y^3, take cube roots, find a value for y, and then a value for x. This is analogous to what happened in Exercise 4.11, where we reduced the problem of finding square roots of $a + bi$ to the problem of solving a quadratic equation with real coefficients. We could solve that equation using square roots of real numbers.

In our current situation, however, it is not clear how to proceed. Indeed, we seem to have made a circle. We wish to compute cube roots of complex numbers in order to solve cubic equations involving real numbers. Now we find that computing cube roots of complex numbers leads us to the problem of solving cubic equations with real number coefficients, and this may lead us back to the calculation of the cube root of a complex number. We will return to this issue in Section 5.5.

4.4 The Complex Plane

We discovered at the end of Section 4.3 that an algebraic approach to calculating cube roots of complex numbers does not seem to work, since it leads to the problem of finding roots of a cubic polynomial with real coefficients, which leads back to the problem of computing cube roots of complex numbers. We need a new idea.

There is such an idea, one that will allow us to calculate not just cube roots but nth roots of complex numbers for any positive integer n, provided we supplement algebra with trigonometry. The idea depends on a geometric interpretation of complex numbers, which we develop in this section.

A complex number $a + bi$ is encoded by the two real numbers a and b, its real and imaginary parts. This allows us to identify $a + bi$ with the pair (a, b). But we can also regard (a, b) as the cartesian coordinates of a point on the plane: a, as usual, is the x-coordinate, specifying how far the point is to the left or right of the y-axis, and b is the y-coordinate, specifying how far the point is above or below the x-axis. For example, we can identify the complex number $1 + \sqrt{3}i$ with the pair $(1, \sqrt{3})$, which we think of as the coordinates of the point on the plane depicted in Figure 4.1.

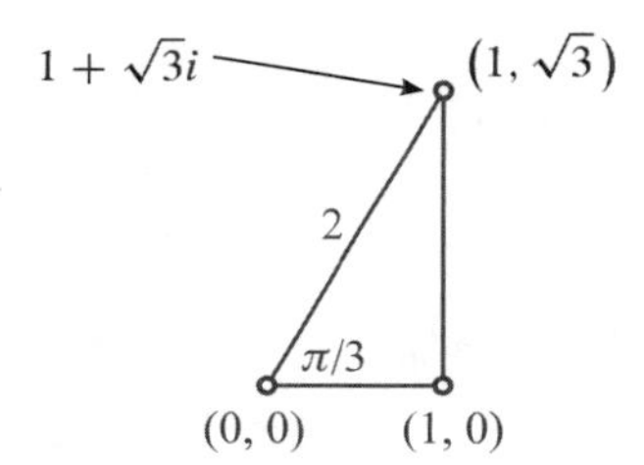

Figure 4.1. The Point $1 + \sqrt{3}i$ in the Plane

Points in the plane can also be specified by their polar coordinates. The *polar coordinates* of a point p in the plane are numbers r and θ, with r representing the distance from p to the origin of the plane and θ representing the angle formed by the positive ray of the x-axis and the line connecting p to the origin. The angle θ is measured in radians, proceeding counterclockwise from the x-axis to the line with p on it. Negative angles indicate that we proceed in a clockwise direction from the x-axis.

Let's write a pair of polar coordinates as $[r, \theta]$, using the brackets to indicate that the pair of numbers is not the usual cartesian coordinates. For example, the point $(1, \sqrt{3})$ (in cartesian coordinates) can be written in polar coordinates as $[2, \pi/3]$. See Figure 4.1.

Polar coordinates have some disadvantages. The principal one is that a point p is described by infinitely many different pairs of polar coordinates. If we move around the plane along a circle, tracing an angle of 2π, we return to the point at which we started. Thus, in addition to $[r, \theta]$, the point p also has polar coordinates $[r, \theta + 2\pi]$. One more trip around the circle and we return to the same point, but this time with polar coordinates $[r, \theta + 4\pi]$. More generally, for any integer n, the point has polar coordinates $[r, \theta + 2n\pi]$. The origin can be described in still more ways, since it has polar coordinates $[0, \theta]$ for arbitrary θ.

Polar coordinates have advantages as well (or else we would not introduce them). One advantage is that certain figures in the plane can be described as the graphs of equations that take on an especially simple form in polar coordinates. For example, $r = 1$ is the equation in polar coordinates of a circle of radius 1 with the origin as its center.

In cartesian coordinates, the circle is described by the equation $x^2 + y^2 = 1$, and we can use it to define the cosine and sine functions for arbitrary real numbers θ. Given a point p on the circle such that the radius from the origin to p forms an angle θ with the x-axis (measured counterclockwise from the x-axis to the radius), the x and y coordinates of p are the cosine and sine of θ:

$$p = (\cos\theta, \sin\theta).$$

This is the key to relating cartesian and polar coordinates.

Exercise 4.15. Let r be a non-negative real number.

(i) The equation of the circle with radius r and center the origin is $x^2 + y^2 = r^2$. Check that a point q on the circle for which the radius from the origin to q forms an angle θ with the x-axis (measured counterclockwise from the x-axis) has cartesian coordinates

$$(r\cos\theta, r\sin\theta).$$

(ii) Conclude that a point with cartesian coordinates (a, b) and polar coordinates $[r, \theta]$ satisfies

$$a = r\cos\theta; \quad b = r\sin\theta.$$

These equations show how to recover the cartesian coordinates of a point from its polar coordinates.

To describe the polar coordinates of a point in terms of its cartesian coordinates, we use inverse trigonometric functions. Any one will do. Let's choose the inverse cosine function, $\arccos t$.

Given a real number t between -1 and 1, there are infinitely many angles θ for which $\cos\theta = t$. Thus, there are infinitely many possibilities for

$\arccos t$. In defining $\arccos t$, by convention we regard θ as restricted to the interval $[0, \pi]$. Within this interval, as θ increases from 0 to π, its cosine will decrease from 1 to -1, assuming all possible values in this range exactly once. The inverse function $\arccos t$ is then defined, for t in $[-1, 1]$, as the unique angle θ between 0 and π for which $\cos\theta = t$. It will decrease from π to 0 as t increases from -1 to 1. It is positive for $t < 1$ and zero for $t = 1$.

Exercise 4.16. Let a and b be real numbers. We wish to find polar coordinates $[r, \theta]$ for the point described in cartesian coordinates by (a, b).

(i) If (a, b) has polar coordinates $[r, \theta]$, explain why $(a, -b)$ has polar coordinates $[r, -\theta]$.

(ii) Describe r in terms of a, b, and square roots.

(iii) Assume $b \geq 0$. Describe θ in terms of a, b, and the inverse cosine function.

(iv) What is θ if $b < 0$?

Exercise 4.17. Let a and b be real numbers.

(i) Using the identification of the complex number $a + bi$ with the point (a, b) in the plane, show that a non-zero complex number $a + bi$ can be written as

$$r\cos\theta + (r\sin\theta)i$$

for some non-negative real number r and some real number θ.

(ii) Write the following complex numbers in this form:

(a) 1 and -1.

(b) i and $-i$.

(c) The four complex numbers $\pm 1 \pm i$.

(d) The four complex numbers $\pm 1 \pm \sqrt{3}i$. (See Figure 4.1.)

We have found that any complex number $a + bi$ can be written as

$$r\cos\theta + (r\sin\theta)i$$

for real numbers r and θ, with r and θ chosen to be polar coordinates of the point with cartesian coordinates (a, b). Exercise 4.16 shows how to express r and θ explicitly in terms of a and b. We call r the *length* or *magnitude* of the complex number $a + bi$ and θ the *angle of inclination* of $a + bi$, since it represents the angle at which the line through (a, b) and $(0, 0)$ inclines with respect to the x-axis. Often, θ is called the *argument* of $a + bi$.

Recall that in Exercise 4.1 we proved that real numbers commute with i under multiplication. Thus, the complex number $a + bi$ is the same as $a + ib$. Sometimes, when working with complicated expressions for the imaginary part, the $a + ib$ form is preferable, in that it eliminates the need for extra parentheses. Thus, for example, we might prefer $r \cos\theta + ir \sin\theta$ to $r \cos\theta + (r \sin\theta)i$. Sometimes it's convenient to write $r \cos\theta + ri \sin\theta$. For example, we might write $2\cos(\pi/3) + 2i \sin(\pi/3)$ rather than putting the i before the 2 or after the sine term.

4.5 A Geometric Interpretation of Multiplication

Having identified complex numbers with points on the plane, we will see in this section how to interpret multiplication of complex numbers geometrically. Let's first review how to interpret multiplication of real numbers geometrically. We identify real numbers with points on the line in the usual way and wish to understand what multiplication by a non-zero real number r does to the line. Begin with three special cases.

If $r = 2$, multiplication of a real number a on the line by 2 yields the number $2a$, which is twice as far from 0 as a is. Thus, multiplication by 2 "expands" the line by a factor of 2. If $r = 1/2$, multiplication of a by $1/2$ yields the number $a/2$, which is half as far from 0 as a is. Thus, multiplication by $1/2$ "contracts" the line by a factor of 2. If $r = -1$, multiplication of a by -1 yields the number $-a$, which is the mirror image of a on the opposite side of the line from the origin, 0. Thus, multiplication by -1 "reflects" the line across 0.

In general, given a positive real number r, we can interpret multiplication by r as an expansion of the real number line if $r > 1$, as a contraction if $r < 1$, and as leaving the line unchanged if $r = 1$. Multiplication by $-r$ performs the same scale change on the line as well as reflecting it across 0.

Multiplication of complex numbers by a non-zero real number has a similar geometric interpretation, but with the real line replaced by the plane. We identify a complex number $a + bi$ with the point (a, b) on the plane. Given a positive real number r, multiplication of the complex number $a+bi$ by r yields $ra + rbi$, which lies on the line through 0 and $a + bi$ in the plane but is r times as far from the origin as $a + bi$. Thus, multiplication by r is an expansion of the plane if $r > 1$, a contraction if $r < 1$, and a fixing of the points on the plane if $r = 1$.

Let's see what multiplication by -1 does to the plane. Suppose $[s, \theta]$ are polar coordinates for the point (a, b), so that

$$a + bi = s \cos \theta + si \sin \theta.$$

Using the trigonometric identities $\cos(\theta + \pi) = -\cos \theta$ and $\sin(\theta + \pi) = -\sin \theta$, we obtain

$$-(a + bi) = s \cos(\theta + \pi) + si \sin(\theta + \pi).$$

Thus, multiplication by -1 leaves the length of each point in the plane unchanged and adds π to its angle of inclination. This results in a rotation of the plane around the origin by π, or 180°. Given a positive real number r, it follows that multiplication by $-r$ is an expansion or contraction of the plane followed by a 180° rotation.

We wish to extend this analysis to obtain a geometric description of multiplication of complex numbers by an arbitrary complex number. The essential case to handle is that of multiplication by i, the most basic of non-real complex numbers. First we check that multiplication of complex numbers satisfies the associative law.

Exercise 4.18. Verify that multiplication of complex numbers is *associative*; that is, given complex numbers p, q, and r, verify that

$$(pq)r = p(qr).$$

To do this, introduce real numbers a, b, c, d, e, and f with $p = a + bi$, $q = c+di$, and $r = e+fi$, then carry out the multiplications and compare, using the associative law for multiplication of real numbers. (The fact that multiplication of real numbers is associative requires proof, but we will take it as a known property of the reals.)

Thanks to associativity, whatever action multiplication by i performs on the plane, performing it twice must produce the same action as multiplication by i^2. The action of i on a complex number p followed by a second action of i yields $i \cdot (i \cdot p)$, whereas the action of i^2 on p is given by $(i^2) \cdot p$, and they are equal. But $i^2 = -1$ and we just saw that multiplication by -1 is a 180° rotation. Thus, performing the action of i twice in succession produces a 180° rotation. This suggests a candidate for the action of multiplication by i on the complex plane: rotation by 90°. There are two choices, rotation clockwise and rotation counterclockwise. That's good, since there are two square roots of -1, the imaginary numbers i and $-i$. We might imagine that one rotation corresponds to multiplication by i, the other to multiplication by $-i$. Let's sort this out in the next exercise.

Exercise 4.19. Let's identify a complex number $a + bi$ with the point (a, b) on the plane and let $[s, \theta]$ be its polar coordinates.

(i) Using polar coordinates and the trigonometric identities

$$\cos(\theta + \pi/2) = -\sin\theta, \quad \sin(\theta + \pi/2) = \cos\theta,$$

verify that

$$\begin{aligned} i(a + bi) &= i(s\cos\theta + is\sin\theta) \\ &= -s\sin\theta + is\cos\theta \\ &= s\cos(\theta + \pi/2) + is\sin(\theta + \pi/2). \end{aligned}$$

(ii) Conclude that multiplication by i rotates the complex plane counterclockwise 90°.

(iii) Similarly, verify that

$$\begin{aligned} -i(a + bi) &= -i(s\cos\theta + is\sin\theta) \\ &= s\sin\theta - is\cos\theta \\ &= s\cos(\theta - \pi/2) + is\sin(\theta - \pi/2) \end{aligned}$$

and conclude that multiplication by $-i$ rotates the complex plane clockwise by 90° (or counterclockwise by 270°).

We introduced complex numbers as formal entities in order to deal with the lack of a square root of -1, declaring i to be such a square root. Additional numbers were introduced so the ordinary rules of addition and multiplication hold, thereby bringing the complex numbers into existence. No concrete meaning was attached to i. It was indeed imaginary. Now, however, having identified complex numbers with points on the plane and multiplication by non-zero real numbers with geometric actions on the plane, we have found a meaning for i. No longer imaginary, it is revealed to be the number that, when used to multiply other complex numbers, produces a counterclockwise rotation by 90°.

To complete our picture, we next need to extend our interpretation of multiplication by i (and $-i$) to multiplication by any complex numbers of length 1. Geometrically, these are the numbers that lie on the unit circle centered at the origin. The angle of inclination of i is 90° and the angle of inclination of $-i$ is 270°. We have seen that multiplication by i and $-i$ produces rotations of exactly those angles. It is not difficult to guess the general result.

Theorem 4.8. *Let u be a complex number with length 1 and angle of inclination θ, so that*

$$u = \cos\theta + i\sin\theta.$$

Let v be a complex number with length s and angle of inclination ϕ, so that

$$v = s\cos\phi + si\sin\phi.$$

Then the product uv has length s and angle of inclination $\theta + \phi$:

$$uv = s\cos(\theta + \phi) + si\sin(\theta + \phi).$$

Thus, multiplication by u effects a rotation of the complex plane counterclockwise through the angle θ.

Theorem 4.8 will follow from the standard trigonometric identities for the sine and cosine of the sum of two angles. These are sufficiently important that we should record them.

Theorem 4.9. *Let θ and ϕ be two real numbers. Then*

$$\sin(\theta + \phi) = \sin\theta\cos\phi + \cos\theta\sin\phi$$

and

$$\cos(\theta + \phi) = \cos\theta\cos\phi - \sin\theta\sin\phi.$$

We should also take a moment to review how Theorem 4.9 is proved, to ensure that we avoid any circularity in our reasoning. Some books provide a proof of Theorem 4.9 as a consequence of de Moivre's formula, which we will prove in Section 4.6 as a consequence of Theorem 4.9. This is the circularity we wish to avoid.

Fortunately, there are many other ways to prove Theorem 4.9. See Chapter 6 of Eli Maor's *Trigonometric Delights* [40, pp. 87–94] for one based on the classical result of plane geometry known as Ptolemy's theorem. This states, for a quadrilateral inscribed in a circle, that the product of the lengths of the quadrilateral's two diagonals equals the sum of the products of lengths of the two pairs of opposite sides. (Ptolemy is the great Alexandrian astronomer and mathematician of the second century C.E.)

Exercise 4.20. Use Theorem 4.9 to show that

$$(\cos\theta + i\sin\theta)\cdot(\cos\phi + i\sin\phi) = \cos(\theta + \phi) + i\sin(\theta + \phi).$$

Deduce that Theorem 4.8 holds.

Let's do some calculations to illustrate Theorem 4.8.

Exercise 4.21. Compute the product of each pair of numbers:

(i) $\cos(\pi/2) + i \sin(\pi/2)$ and $\cos(3\pi/2) + i \sin(3\pi/2)$.

(ii) $\cos(\pi/4) + i \sin(\pi/4)$ and $\cos(\pi/4) + i \sin(\pi/4)$.

(iii) $\cos(\pi/6) + i \sin(\pi/6)$ and $\cos(\pi/3) + i \sin(\pi/3)$.

(iv) $\cos(\pi/18) + i \sin(\pi/18)$ and $\cos(\pi/5) + i \sin(\pi/5)$.

For the first three examples, rewrite the numbers in more familiar terms, using the explicit values of the trigonometric functions.

We conclude the sequence of ideas in this section with a description of the product of two complex numbers.

Exercise 4.22. Suppose we are given a complex number u with length r and angle of inclination θ and a complex number v with length s and angle of inclination ϕ. Thus

$$u = r \cos\theta + ri \sin\theta$$

and

$$v = s \cos\phi + si \sin\phi.$$

Show that

$$(r \cos\theta + ri \sin\theta) \cdot (s \cos\phi + si \sin\phi) = rs \cos(\theta + \phi) + rs\, i \sin(\theta + \phi).$$

Reinterpret this to say that the product uv is the complex number whose length is the product rs of the lengths r and s of u and v and whose angle of inclination $\theta + \phi$ is the sum of the angles of inclination θ and ϕ of u and v.

We have found that multiplying two complex numbers amounts, geometrically, to multiplying their lengths and adding their angles of inclination.

4.6 Euler's and de Moivre's Formulas

Our motivation in studying complex numbers is our desire to compute their cube roots. The result that will allow us to do this—*de Moivre's formula*—provides at the same time a procedure for computing nth roots for any positive integer n. (The formula's name honors Abraham de Moivre (1667–1754), a French mathematician who spent much of his life in England.)

Thus, we may as well consider the general case. However, we will lay out the issues in such a way that the reader can ignore the general case and focus on the cases $n = 2$ and $n = 3$.

One way to derive de Moivre's formula is to use trigonometry and Theorem 4.9. We will take this approach in a moment. But first let's see how easily it follows from Euler's formula, a result due to Leonhard Euler (1707–1783), the greatest mathematician of the eighteenth century.

Euler's formula involves the number e, whose central role in mathematics Euler first discovered and whose notation he introduced. Readers who have studied calculus will be familiar with it as the number most naturally used for exponentiation. What makes it so natural is that the exponential function e^x with base e is its own derivative, in contrast to other exponential functions such as 2^x and 10^x. Another attractive feature is that its inverse function $\log_e x$ is the antiderivative of $1/x$. For readers unfamiliar with e, suffice to say that it is an irrational number with decimal expansion that begins 2.71828 and that for theoretical reasons it is the best base for studying exponentiation.

Euler's formula relates exponential and trigonometric functions:

Theorem 4.10 (Euler's formula). *Let θ be a real number. Then*

$$e^{i\theta} = \cos\theta + i\sin\theta.$$

At first glance, the formula may appear both mysterious and wondrous. For those readers familiar with power series expansions of functions, the mystery is easily addressed, but the wonder remains.

In a calculus course, it is shown that the sine and cosine functions have power series expansions

$$\cos x = 1 - \frac{x^2}{2!} + \frac{x^4}{4!} - \frac{x^6}{6!} + \cdots$$

and

$$\sin x = x - \frac{x^3}{3!} + \frac{x^5}{5!} - \frac{x^7}{7!} + \cdots,$$

and the exponential function satisfies

$$e^x = 1 + \frac{x}{1!} + \frac{x^2}{2!} + \frac{x^3}{3!} + \frac{x^4}{4!} + \frac{x^5}{5!} + \cdots.$$

The expansions are valid for any real number x.

Once complex numbers are introduced, it is not a large leap to replace real numbers in the power series expansion of e^x with pure imaginary numbers. Suppose for instance that θ is a real number and let $x = i\theta$. If we substitute $i\theta$ for x in the power series expansion of e^x and use the equalities $i^2 = -1$, $i^3 = -i$, and $i^4 = 1$, we find that

$$e^{i\theta} = 1 + i\theta - \frac{\theta^2}{2!} - i\frac{\theta^3}{3!} + \frac{\theta^4}{4!} + i\frac{\theta^5}{5!} + \cdots.$$

Collecting real and imaginary terms on the right side of this equation, we obtain $\cos\theta + i\sin\theta$, yielding Euler's formula. That's all there is to it.

Given a real number θ and a positive integer n, the rules of exponentiation yield

$$(e^{i\theta})^n = e^{in\theta}.$$

Reinterpreting the left and right sides of this equation via Euler's formula, we see that

$$(\cos\theta + i\sin\theta)^n = \cos n\theta + i\sin n\theta.$$

This is de Moivre's formula:

Theorem 4.11 (de Moivre's formula)**.** *Let n be a positive integer and θ a real number. Then*

$$(\cos\theta + i\sin\theta)^n = \cos n\theta + i\sin n\theta.$$

In words, de Moivre's formula says that the nth power of a complex number of length 1 is the number obtained by multiplying the original number's angle of inclination by n. In particular, squaring amounts to doubling the angle and cubing amounts to tripling the angle.

We were led to de Moivre's formula from Euler's formula, but the proof of Euler's formula requires ideas from calculus. An alternative, more elementary proof of de Moivre's formula is desirable. As de Moivre understood, the underlying issue is one of trigonometry. The proof sketched in Exercise 4.23 uses the classical angle sum formulas of Theorem 4.9 (and an implicit induction argument). The cases $n = 2$ and $n = 3$, which are all we need, are treated separately before we handle the general case.

Exercise 4.23. Prove de Moivre's formula.

(i) The formula holds trivially for $n = 1$.

(ii) Square $\cos\theta + i\sin\theta$, combine real and imaginary terms, and use Theorem 4.9 to show that

$$(\cos\theta + ri\sin\theta)^2 = \cos 2\theta + i\sin 2\theta.$$

(iii) Multiply both sides of the last equality by $\cos\theta + i\sin\theta$ and use Theorem 4.9 to show that

$$(\cos\theta + ri\sin\theta)^3 = \cos 3\theta + i\sin 3\theta.$$

(iv) We can continue this process step by step to obtain the general result. Suppose k is a positive integer less than n and we know that

$$(\cos\theta + i\sin\theta)^k = \cos k\theta + \ i\sin k\theta.$$

Multiply both sides of this equality by $\cos\theta + i\sin\theta$ and use Theorem 4.9 to deduce that

$$(\cos\theta + i\sin\theta)^{k+1} = \cos(k+1)\theta + \ i\sin(k+1)\theta.$$

(v) Deduce that we can make this argument $n-1$ times successively to obtain the formula for $k = 2$, then $k = 3$, then $k = 4$, and so on, until we obtain the formula for $k = n$.

Exercise 4.24. Let's do some computations using de Moivre's formula.

(i) Compute

$$(\cos(\pi/4) + i\sin(\pi/4))^n$$

for $n = 2, 3, 4$, and 8, thus finding

$$\left(\frac{\sqrt{2}}{2} + \frac{\sqrt{2}}{2}i\right)^n$$

for these values of n.

(ii) Compute

$$\left(\frac{1}{2} + \frac{\sqrt{3}}{2}i\right)^n$$

for $n = 2, 3$, and 6.

Exercise 4.25. As a minor extension of de Moivre's formula, prove Theorem 4.12, which describes the nth power of a non-zero complex number.

Theorem 4.12. *Let n be a positive integer, r a positive real number, and θ a real number. Then*

$$(r\cos\theta + ri\sin\theta)^n = r^n\cos n\theta + r^n i\sin n\theta.$$

De Moivre's formula and Theorem 4.12 allow us to compute nth powers of complex numbers instantly, once the numbers are expressed in the form $r\cos\theta + ri\sin\theta$. They also reveal how to calculate nth roots: we do the reverse, calculating the real nth root of the length of a complex number and dividing the angle of inclination by n.

The procedure has the drawback that it is not purely algebraic. It requires trigonometry to determine the angle of inclination of a given complex number and trigonometry again, once nth roots have been taken, to calculate the real and imaginary parts.

Let's work through the details, dealing first with square roots and cube roots. We already found how to compute the two square roots of a non-zero complex number in Exercise 4.11, using algebra, but now we'll use the trigonometric approach.

Exercise 4.26. Suppose we wish to compute the square root of a non-zero complex number c. Write c as

$$r\cos\theta + ri\sin\theta$$

for some positive real number r and some real θ. Write $\sqrt{r}$ for the positive real square root of r.

(i) Using Theorem 4.12, verify that the two complex numbers

$$\sqrt{r}\,\cos\frac{\theta}{2} + \sqrt{r}\,i\,\sin\frac{\theta}{2}$$

and

$$\sqrt{r}\,\cos(\frac{\theta}{2} + \pi) + \sqrt{r}\,i\,\sin(\frac{\theta}{2} + \pi)$$

are square roots of c.

(ii) Check that each is the opposite of the other, as we would expect.

(iii) Use the formula to compute square roots of

(a) 4 and -4.

(b) i and $-i$.

(c) $1+i$ and $-1+i$.

Next let's find cube roots.

Exercise 4.27. Suppose c is a non-zero complex number. Write it as

$$r\cos\theta + ri\sin\theta$$

for some positive real number r and some real θ.

(i) Recall that r has a unique real cube root. Write it as $\sqrt[3]{r}$.

(ii) Using Theorem 4.12, show that

$$\sqrt[3]{r}\cos\frac{\theta}{3} + \sqrt[3]{r}\, i\sin\frac{\theta}{3}$$

is a cube root of c, as are

$$\sqrt[3]{r}\cos\left(\frac{\theta}{3} + \frac{2\pi}{3}\right) + \sqrt[3]{r}\, i\sin\left(\frac{\theta}{3} + \frac{2\pi}{3}\right)$$

and

$$\sqrt[3]{r}\cos\left(\frac{\theta}{3} + \frac{4\pi}{3}\right) + \sqrt[3]{r}\, i\sin\left(\frac{\theta}{3} + \frac{4\pi}{3}\right).$$

(iii) As in Exercise 3.15, let

$$\omega = -\frac{1}{2} + \frac{\sqrt{-3}}{2}.$$

Show, using the formulas, that the three cube roots of 1 are 1, ω, and ω^2. (Hint: First express 1 as $\cos 0 + i\sin 0$.)

(iv) Suppose d is one of the cube roots of c. Verify that the other two are ωd and $\omega^2 d$.

(v) Suppose c is a non-zero real number, with real cube root d. Conclude that c has one real cube root, d, and two non-real cube roots, ωd and $\omega^2 d$.

We can use Exercise 4.27 to redo some cube root calculations that we made earlier when we used Cardano's formula, as illustrated next.

Exercise 4.28. In Exercise 3.13, we found a cube root of

$$-3 + \frac{10}{9}\sqrt{-3}$$

and a cube root of

$$-3 - \frac{10}{9}\sqrt{-3}.$$

Let's now find all three cube roots.

(i) Use the method of Exercise 4.27 and a calculator to find a cube root of $-3 + 10\sqrt{-3}/9$. Then use Exercise 4.27 to find its other two cube roots.

(ii) In the same way, find the three cube roots of $-3 - 10\sqrt{-3}/9$.

(iii) Alternatively, use Theorem 4.1 to find its three cube roots.

(iv) Compare the results to those arising from the calculation of cube roots in Exercise 3.13.

We have focused so far on finding square and cube roots of a non-zero complex number, but the same approach allows us to find nth roots for any positive integer n. As a consequence of Theorem 1.12, any positive real number has a unique positive real nth root, which we write $\sqrt[n]{r}$. Here is the general result:

Theorem 4.13. *Let r be a positive real number. The complex number*

$$r \cos\theta + ri \sin\theta$$

has n distinct nth roots, given by

$$\sqrt[n]{r} \cos\left(\frac{\theta}{n} + \frac{2\pi j}{n}\right) + \sqrt[n]{r}\, i \sin\left(\frac{\theta}{n} + \frac{2\pi j}{n}\right)$$

as $j = 0, 1, \ldots, n-1$.

Exercise 4.29. Prove Theorem 4.13:

(i) Verify that the n quantities are distinct.

(ii) Verify that they are nth roots.

4.7 Roots of Unity

The only result from this section that we will use later is Theorem 4.16, which can be proved by direct calculation. Thus, the reader may wish to go straight to Theorem 4.16, and its proof, and move ahead. However, a natural generalization of Theorem 4.16 is easy to obtain, and it enhances our understanding of nth roots.

A complex number that has an nth power equal to 1, where n is a positive integer, is called a *root of unity*. Thanks to Theorem 4.13 (or de Moivre's formula), we have identified them all.

Theorem 4.14. *The n distinct nth roots of unity are*

$$\cos\left(\frac{2\pi j}{n}\right) + i \sin\left(\frac{2\pi j}{n}\right),$$

as $j = 0, 1, \ldots, n-1$.

Exercise 4.30. Prove Theorem 4.14.

Exercise 4.31. For small n, we can write down the nth roots of unity as ordinary complex numbers, without trigonometric functions in their expressions. We have already done so in several cases. Let's review and extend what we know. Describe without cosines and sines the two square roots of 1, the three cube roots of 1, the four fourth roots of 1, the six sixth roots of 1, and the eight eighth roots of 1.

The two square roots of 1 are the powers of -1, the three cube roots of 1 are the powers of the number we have called ω, and the four fourth roots of 1 are the powers of i. For an arbitrary positive integer n, let ζ_n be the number

$$\cos\frac{2\pi}{n} + i\sin\frac{2\pi}{n}.$$

Here, ζ is the Greek letter zeta. Notice that ζ_2 is -1 and ζ_3 is ω, while ζ_4 is i.

We learn different parts of Theorem 4.15 early in our mathematical education. One might not think to state it as a theorem, but it provides the template for the results that follow.

Theorem 4.15. *Let* ζ_2 *be the number* $\cos(2\pi/2) + i\sin(2\pi/2)$. *For simplicity, write* ζ_2 *in its more familiar form as* -1.

(i) The two distinct square roots of 1 *are* 1 *and* -1, *with* $(-1)^2 = 1$.

(ii) $1 + (-1) = 0$.

(iii) $x^2 - 1 = (x-1)(x-(-1))$.

(iv) $A^2 - B^2 = (A-B)(A-(-1)B)$.

Theorem 4.15 has an analogue for any integer $n > 2$, with 1 and -1 replaced by the nth roots of unity. We will need the appropriate analogue for $n = 3$ in Section 5.3.

Theorem 4.16. *Let* ζ_3 *be the number* $\cos(2\pi/3) + i\sin(2\pi/3)$. *For simplicity, write* ζ_3 *in its more familiar form as* ω.

(i) The three distinct cube roots of 1 *are* 1, ω, *and* ω^2, *with* $\omega^3 = 1$.

(ii) $1 + \omega + \omega^2 = 0$.

(iii) $x^3 - 1 = (x-1)(x-\omega)(x-\omega^2)$.

(iv) $A^3 - B^3 = (A-B)(A-\omega B)(A-\omega^2 B)$.

Exercise 4.32. Prove Theorem 4.16. The first part has already been shown in Exercise 3.15 (or Exercise 4.31). Verify the second part by direct calculation, and use it to prove the third and fourth parts.

The analogous result for $n = 4$ is even easier to prove.

Theorem 4.17. *Let ζ_4 be the number $\cos(2\pi/4) + i\sin(2\pi/4)$. For simplicity, write ζ_4 in its more familiar form as i.*

(i) The four distinct fourth roots of 1 *are* $1, i, i^2,$ *and* i^3*, with* $i^4 = 1$.

(ii) $1 + i + i^2 + i^3 = 0$.

(iii) $x^4 - 1 = (x-1)(x-i)(x-i^2)(x-i^3) = (x-1)(x+1)(x-i)(x+i)$.

(iv) $$\begin{aligned} A^4 - B^4 &= (A-B)(A-iB)(A-i^2B)(A-i^3B) \\ &= (A-B)(A+B)(A-iB)(A+iB). \end{aligned}$$

Exercise 4.33. Prove Theorem 4.17.

Here is the theorem that generalizes Theorems 4.15, 4.16, and 4.17:

Theorem 4.18. *Let n be a positive integer and let ζ_n be the number $\cos(2\pi/n) + i\sin(2\pi/n)$.*

(i) The n distinct nth roots of unity are $1, \zeta_n, \zeta_n^2, \ldots, \zeta_n^{n-1}$*, with* $(\zeta_n)^n = 1$.

(ii) $1 + \zeta_n + \zeta_n^2 + \cdots + \zeta_n^{n-2} + \zeta_n^{n-1} = 0$.

(iii) $x^n - 1 = (x-1)(x-\zeta_n)(x-\zeta_n^2) \cdots (x-\zeta_n^{n-1})$.

(iv) $A^n - B^n = (A-B)(A-\zeta_n B)(A-\zeta_n^2 B)\cdots(A-\zeta_n^{n-1}B)$.

Exercise 4.34. Prove Theorem 4.18.

(i) Use Theorem 4.14 and de Moivre's formula to obtain the first part.

(ii) For the second part, verify that $x^n - 1$ factors as

$$x^n - 1 = (x-1)(x^{n-1} + x^{n-2} + \cdots + x^2 + x + 1).$$

Then substitute ζ_n for x and use the fact that $\zeta_n \neq 1$ to deduce the desired result.

(iii) The proof of Theorem 1.5 works just as well for complex roots of a polynomial as for real roots. Deduce that Theorem 1.5 can be applied to show that

$$(x-1)(x-\zeta_n)(x-\zeta_n^2)\ \cdots\ (x-\zeta_n^{n-1})$$

divides $x^n - 1$. Compare degrees and deduce that $x^n - 1$ must be a constant multiple of the product. Then compare the coefficients of x^n and deduce that equality holds.

(iv) Deduce the fourth part from the third using the substitution $x = A/B$.

4.8 Converting Root Extraction to Division

Throughout our discussion of the quadratic formula and Cardano's formula, we have emphasized that the role of algebra is to reduce the problem of solving quadratic and cubic equations to that of calculating square and cube roots, but that algebra does not provide us with the means to make the root calculations. Let's discuss how we can in principle make the calculations (and more generally nth root calculations) through the use of exponential and trigonometric functions and their inverses. We will once again assume familiarity with the exponential function e^x, where the base e is Euler's number, which we met in Section 4.6. We will also use the inverse function $\log_e(x)$, the *natural logarithm* function, which we'll write as $\log(x)$.

Exercise 4.35. Let r be a positive real number and n a positive integer. Suppose we wish to calculate the real nth root of r.

(i) Set $c = r^{1/n}$, the real nth root. Apply $\log(x)$ to both sides to obtain

$$\log(c) = \log(r)/n.$$

(ii) Exponentiate both sides to obtain

$$c = e^{\log(c)} = e^{\log(r)/n}.$$

(iii) Rewrite this as

$$r^{1/n} = e^{\log(r)/n}.$$

(iv) Conclude that this reduces the problem of calculating nth roots of positive real numbers to that of computing certain values of the exponential and logarithm functions and dividing by n.

The lesson of Exercise 4.35 is that applying the logarithm converts extraction of an nth root of a real number to division by n.

Exercise 4.36. Let a and b be real numbers satisfying $a^2 + b^2 = 1$ and let n be a positive integer. Suppose we wish to calculate an nth root of $a + bi$. For simplicity in the trigonometric considerations to follow, assume that $b \geq 0$. Then $a + bi$ can be written as $\cos \theta + i \sin \theta$ for an angle θ given by

$$\theta = \arccos a.$$

(i) Use Theorem 4.13 to deduce that one nth root of $\cos \theta + i \sin \theta$ is

$$\cos(\theta/n) + i \sin(\theta/n).$$

(ii) By substitution, conclude that one nth root of $a + bi$ is given by

$$\cos((\arccos a)/n) + i \sin((\arccos a)/n).$$

(iii) We may rewrite this in another way. The sine and cosine functions satisfy

$$\sin x = \cos(x - \pi/2).$$

Rewrite the expression for an nth root of $a + bi$ as

$$\cos((\arccos a)/n) + i \cos((\arccos a)/n - \pi/2).$$

(iv) Conclude that this reduces the problem of calculating nth roots of complex numbers of length 1 to that of computing certain values of the inverse cosine and cosine functions and dividing by n.

A non-zero complex number $a+bi$ can be written in the form $r(\cos \theta + i \sin \theta)$ for a positive real number r and a real number θ. Exercises 4.35 and 4.36 give us procedures for computing the nth roots of r and $\cos \theta + i \sin \theta$, allowing us to compute an nth root of $a + bi$.

The calculational methods of the last two exercises have a common feature. In each nth root calculation, we make use of a familiar function $f(x)$ and its inverse $f^{-1}(x)$, applying f^{-1} to a real number c, dividing what we obtain by n, then calculating f on the result. Both times, this process converts nth root extraction to division by n. The cost of this simplification is the requirement that we have the means to compute f^{-1} and f on given real numbers.

4.9 History

Complex numbers are so widely used that it can be difficult to understand why their acceptance took centuries, stretching from their appearance in Cardano's *Ars Magna* [12] in 1545 to the development of the subject now called complex analysis in the late nineteenth century. Two entertaining accounts of the intellectual challenge complex numbers posed are Paul J. Nahin's *An Imaginary Tale: The Story of* $\sqrt{-1}$ [43] and Barry Mazur's *Imagining Numbers (Particularly the Square Root of Minus Fifteen)* [41]. They serve as natural complements to each other, with rich historical discussions that are especially worth reading. In this section, we will touch on just a few historical highlights.

Let's begin with de Moivre's formula. It does not appear that de Moivre wrote the formula down explicitly, but he surely knew it. Or at least he understood that taking nth roots of complex numbers produces formulas similar to trigonometric formulas for sines and cosines, allowing him to relate taking nth roots with dividing angles by n. For instance, in a 1707 note with the title "The analytic solution of certain equations of the third, fifth, seventh, ninth and other higher uneven powers, by rules similar to those called Cardan's," he writes [16], [60, pp. 443–444]:

If the equation were

$$5y - 20y^3 + 16y^5 = \frac{61}{64},$$

$y =$

$$\frac{1}{2}\sqrt[5]{\frac{61}{64} + \sqrt{\frac{-375}{4096}}} + \frac{1}{2}\sqrt[5]{\frac{61}{64} - \sqrt{\frac{-375}{4096}}}.$$

And if by any means the fifth root of the binomial can be extracted the root will come out true and possible, although the expression seems to include an impossibility. Now the fifth root of the binomial

$$\frac{61}{64} + \sqrt{\frac{-375}{4096}}$$

is

$$\frac{1}{4} + \frac{1}{4}\sqrt{-15},$$

and of the binomial

$$\frac{61}{64} - \sqrt{\frac{-375}{4096}}$$

> is
> $$\frac{1}{4} - \frac{1}{4}\sqrt{-15}$$
> whose semi-sum $1/4 = y$. But if the extraction cannot be performed, or should seem too difficult, the thing may always be effected by a table of natural sines in the following manner.
>
> To the radius 1 let $a = 61/64 = 0.95112$ be the sine of a certain arc which therefore will be $72^\circ\ 23'$, the fifth part of which (because $n = 5$) is $14^\circ\ 28'$; the sine of this is $0.24981, nearly = 1/4$. So also for equations of higher degree.

De Moivre recognizes that $(61/64) + i\sqrt{375/4096}$ has the form $\sin\theta + i\cos\theta$ for a suitable θ, and adding the fifth roots of this number and its conjugate amounts to computing the sine of $\theta/5$. As Mazur comments ([41, p. 199]), "Although there were earlier hints of the link between the solution of polynomial equations and trigonometry, we know that Abraham De Moivre, by 1707, had perceived the analogy between the geometric problem of cutting an arc of a circle into n equal parts and taking the nth root of a complex number."

In 1748, Leonhard Euler stated de Moivre's formula in its standard form, deriving it using the angle-sum formulas for sine and cosine. Here is the start of his discussion:

> Since $(\sin z)^2 + (\cos z)^2 = 1$, on decomposing into factors we get
> $$(\cos z + i\sin z)(\cos z - i\sin z) = 1.$$
> These factors, although imaginary, are of great use in the combination and multiplication of arcs. For example, let us seek the product of these factors.

Euler then continues in a familiar way to find the square, the cube, and the nth power of $\cos z \pm i\sin z$.

Euler made many contributions to the understanding of complex numbers, including his famous formula relating the exponential, cosine, and sine functions. Beyond the results themselves, he played an essential role in bringing about the acceptance of their use. For example, in his 1770 text *Elements of Algebra* [26], Euler introduces complex numbers early, in Chapter XIII of Section 1, where he writes [26, pp. 42–43]:

> When it is required, therefore, to extract the root of a negative number, a great difficulty arises; since there is no assignable number, the square of which would be a negative quantity. ... it is evident that we cannot

> rank the square root of a negative number amongst the possible numbers, and we must therefore say that it is an impossible quantity. In this manner we are led to the idea of numbers, which from their nature are impossible; and therefore they are usually called *imaginary quantities*, because they exist merely in the imagination. ... But notwithstanding this, these numbers present themselves to the mind; they exist in our imagination, and we still have a sufficient idea of them; ... for this reason also, nothing prevents us from making use of these imaginary numbers, and employing them in calculation.

Elements of Algebra would become an influential text. A translation from German to French was prepared by Johann III Bernoulli, with about one hundred pages of additions by Joseph-Louis Lagrange. An English translation of the French edition was begun by Francis Horner, who also wrote a short biography of Euler, and completed by Reverend John Hewlett. The 1840 edition of the English translation containing Bernoulli's notes, Lagrange's additions, and Horner's biography remains available thanks to Springer-Verlag.

A good account of Euler's work on complex numbers can be found in Chapter 5 of William Dunham's *Euler: The Master of Us All* [23]. Dunham concludes it with the observation [23, pp. 101–102] that "complex numbers were here to stay. A concept only dimly understood for its role in solving cubic equations had been legitimized by the discoveries and influence of Leonhard Euler. Without apology or embarrassment, he treated these numbers as equal players on the mathematical stage and showed how to take their roots, logs, sines, and cosines."

As the eighteenth century gave way to the nineteenth, the geometric description of complex numbers presented in Sections 4.4 and 4.5 first appeared. How it did is a fascinating story, well told in the books of Nahin [43] and Mazur [41]. We will provide a brief sketch. But first we should note that an earlier effort to describe pure imaginary numbers as points on the plane that lie off the real number line appears in a 1685 work on algebra [69] by the English mathematician John Wallis (1616–1703). His approach was inconclusive. We will return to some of the material in Wallis's book in Section 5.8.

The first person to describe complex numbers geometrically in the way we now understand was the Norwegian Caspar Wessel (1745–1818). He presented a paper to the Royal Danish Academy of Sciences in 1797 that was published in Danish two years later in the Academy's *Memoires* [70]. Wessel recognized that in multiplying two complex numbers, one adds their angles of inclination. It appears, however, that Wessel's work went unnoticed for nearly a century, until it was rediscovered in 1895 [43, pp. 48–49, 243].

In 1806, the Swiss-born Jean-Robert Argand (1768–1822) published a pamphlet, *Essay on the Geometrical Interpretation of Imaginary Quantities*, in which he also identified complex numbers with points on the plane. (We will quote from A.S. Hardy's 1881 translation [4].) Before tackling complex numbers, Argand reviews the one-time challenge of making sense of negative numbers [4, pp. 17–22], concluding that

> the difficulty of the subject will not be questioned if we remember that the exact sciences had been cultivated for many centuries, and had made great progress before either a true conception of negative quantities was reached or a general method for their use had been devised.

The notion of a negative number, Argand explains, might seem imaginary, but when we compare two quantities, we consider not just the ratio of their absolute values, but also "a relation of direction, or of the sense in which they are estimated, a relation either of identity or opposition."

Next Argand poses the problem [4, p. 23] of finding "the geometric mean between two quantities of different signs, that is, to find the value of x in the proportion $+1 : x :: x : -1$." In effect, he asks for a a value of x satisfying $1/x = x/(-1)$, or $x^2 = -1$.

> Here we encounter a difficulty ...; but, as before, the quantity which was imaginary, when applied to certain magnitudes, became real when to the idea of absolute number we added that of direction, may it not be possible to treat this quantity, which is regarded imaginary, because we cannot assign it a place in the scale of positive and negative quantities, with the same success? On reflection this has seemed possible, provided we can devise a kind of quantity to which we may apply the idea of direction, so that having chosen two opposite directions, one for positive and one for negative values, there shall exist a third — such that the positive direction shall stand in the same relation to it that the latter does to the negative.

Having opened the door to additional directions, Argand explains that the quantity x which is to be the geometric mean of 1 and -1 should be perpendicular to the line containing them. This yields two choices, which are "related to each other as $+1$ and -1. They are, therefore, what is ordinarily expressed by $+\sqrt{-1}$ and $-\sqrt{-1}$." So begins Argand's identification of complex numbers with points on the plane.

Argand's work was ignored initially. Guillaume-Jules Hoüel tells the story in his preface to the 1876 reprint of Argand's book [4, pp. iii–xvi]. (See also [43, pp. 73–74].) Argand was living in Paris at the time, and he sent a copy to the great French mathematician Adrien-Marie Legendre. Without mentioning Argand's name, Legendre described Argand's ideas in a letter to François Français, a professor of mathematics. On François's death in 1810, his younger brother Jacques inherited his papers. Jacques published an article in 1813 in the *Annales de Mathématiques* describing Argand's ideas as well as Legendre's letter, and asked who the unknown author of the pamphlet might be. Argand saw Français's article and identified himself, receiving full credit from Français.

Carl Friedrich Gauss (1777–1855), the greatest mathematician of the era, had come up with the same ideas, perhaps even earlier than Wessel. He chose not to publish his work until 1831, yet he came to share naming rights with Argand. The plane, when its points are identified with complex numbers, has come to be called the *Gaussian plane*, *Argand plane*, or *Argand diagram*.

5

Cubic Polynomials, II

We derived Cardano's formula for roots of reduced cubic polynomials in Section 3.2, only to discover that using it may require us to compute cube roots of complex numbers. This phenomenon arose in Exercise 3.12, when we tried to solve the cubic equation $y^3 - 7y + 6 = 0$. Our study of the discriminant in Section 3.4 revealed that this difficulty will occur whenever we work with a cubic whose roots are real and distinct. With that, we brought Chapter 3 to a close and turned to a study of complex numbers in Chapter 4. Now that we have learned how to use trigonometry to compute cube roots of complex numbers, we are ready for a more systematic treatment of cubic polynomials.

5.1 Cardano's formula

Given real numbers p and q, Cardano's formula for a root of the reduced cubic polynomial $y^3 + py + q$ takes the form

$$y = \sqrt[3]{-\frac{q}{2} + \sqrt{\frac{p^3}{27} + \frac{q^2}{4}}} + \sqrt[3]{-\frac{q}{2} - \sqrt{\frac{p^3}{27} + \frac{q^2}{4}}}.$$

Our derivation of the formula in Exercise 3.7 lacked precision, first because we didn't yet know how to interpret the expressions within the cube root signs if $p^3/27 + q^2/4 < 0$ and second because we didn't yet understand that there are three possibilities for each cube root. We can proceed now with confidence.

Let us once again write R for $p^3/27 + q^2/4$ and ω for $-1/2 + \sqrt{-3}/2$, which is a cube root of 1. We learned in Exercise 4.27 how to compute cube roots of $-q/2+\sqrt{R}$. Let A be one of them. Then the other two are ωA and $\omega^2 A$. Similarly, if B is a cube root of $-q/2-\sqrt{R}$, then the other two are ωB and $\omega^2 B$. In the next exercise, we determine which cube roots to add together in order to obtain roots of y^3+py+q.

Exercise 5.1. Let p and q be real numbers with $p \neq 0$. (We can handle the $p = 0$ case directly.) Since any non-zero number has three distinct cube roots, each summand on the right side of Cardano's formula has three possible values.

(i) Check, as in Exercise 3.7, that the product of $-q/2+\sqrt{R}$ and $-q/2-\sqrt{R}$ is $-p^3/27$.

(ii) The derivation in Exercise 3.7 showed that for the sum of the cube roots of $-q/2+\sqrt{R}$ and $-q/2-\sqrt{R}$ to be a root of y^3+py+q, the two cube roots must be chosen so that their product is $-p/3$. Select one of the three cube roots of $-q/2+\sqrt{R}$—any one, it doesn't matter which—and call it A.

(iii) Although A is chosen arbitrarily, we will exercise care in choosing a cube root B of $-q/2-\sqrt{R}$ so that $A+B$ is a root of y^3+py+q. We have three choices for it. We will want to choose a specific one, given the choice of A. For now, let us choose one arbitrarily and call it $\mathcal{B}$. The other two are $\omega\mathcal{B}$ and $\omega^2\mathcal{B}$.

(iv) Check that $A^3\mathcal{B}^3 = -p^3/27$ and deduce that $A\mathcal{B}$ equals one of the numbers $-p/3$, $-\omega p/3$, and $-\omega^2 p/3$.

(v) If $A\mathcal{B} = -p/3$, set $B = \mathcal{B}$. If $A\mathcal{B} = -\omega p/3$, set $B = \omega^2\mathcal{B}$; and if $A\mathcal{B} = -\omega^2 p/3$, set $B = \omega\mathcal{B}$. Verify in all cases that $AB = -p/3$.

(vi) Show, with A and B chosen in this way, that the numbers ωA and $\omega^2 B$ are cube roots of $-q/2+\sqrt{R}$ and $-q/2-\sqrt{R}$ with the property that their product equals $-p/3$. Show also that the numbers $\omega^2 A$ and ωB are cube roots of $-q/2+\sqrt{R}$ and $-q/2-\sqrt{R}$ with the property that their product equals $-p/3$.

(vii) Conclude that if A and B are chosen as cube roots of $-q/2+\sqrt{R}$ and $-q/2-\sqrt{R}$ satisfying $AB = -p/3$, then the three roots of y^3+py+q have the form

$$r_1 = A+B;\quad r_2 = \omega A+\omega^2 B;\quad r_3 = \omega^2 A+\omega B.$$

Exercise 5.1 removes the imprecision that was present in our earlier derivation of Cardano's formula. We can summarize what we have found in a theorem.

Theorem 5.1 (Cardano's formula)**.** *Suppose p and q are real numbers, with $p \neq 0$. Let A be a cube root of*

$$-\frac{q}{2} + \sqrt{\frac{p^3}{27} + \frac{q^2}{4}}$$

and let B be the unique cube root of

$$-\frac{q}{2} - \sqrt{\frac{p^3}{27} + \frac{q^2}{4}}$$

satisfying $AB = -p/3$. Let ω be the cube root $-1/2 + \sqrt{-3}/2$ of 1. Then the three roots of the polynomial $y^3 + py + q$ are

$$A + B,\ \omega A + \omega^2 B,\ \omega^2 A + \omega B.$$

Exercise 5.2. Use Cardano's formula, as clarified in Theorem 5.1, to solve the cubic equation

$$y^3 + 6y - 20 = 0$$

that we first studied in Exercises 3.9 and 3.10. There's no need to redo work already done. Use the result of Exercise 3.10 to write a cube root for each term in the formula. Then use what we have learned since then to write two more cube roots for each term, pair them correctly, and take their sums to find the three solutions to the equation. One solution is real and two are not.

Given a cubic polynomial $y^3 + py + q$ for which R is negative, the numbers $-q/2 + \sqrt{R}$ and $-q/2 - \sqrt{R}$ will not be real, and the calculation of the roots given by Cardano's formula requires us to compute cube roots of non-real numbers.

Exercise 5.3. Use Cardano's formula to solve the cubic equation

$$y^3 - 7y + 6 = 0$$

that we first studied in Exercise 3.13.

(i) Write the solution given by the formula as a sum of cube roots. It involves the cube roots of

$$-3 + \frac{10}{9}\sqrt{-3}$$

and

$$-3 - \frac{10}{9}\sqrt{-3}.$$

(ii) Using the numbers ω and ω^2 and the determination of one cube root of $-3 + \frac{10}{9}\sqrt{-3}$ in Exercise 3.13 or Exercise 4.28, write expressions for the three complex numbers that are cube roots of $-3 + \frac{10}{9}\sqrt{-3}$. Write also the three complex numbers that are cube roots of $-3 - \frac{10}{9}\sqrt{-3}$.

(iii) Pair the cube roots of $-3 + \frac{10}{9}\sqrt{-3}$ and $-3 - \frac{10}{9}\sqrt{-3}$ as specified in Theorem 5.1 to get three pairs such that the product of the complex numbers in each pair equals 7/3.

(iv) Add the complex numbers in each pair to obtain all three real solutions of $y^3 - 7y + 6 = 0$.

When making calculations of cube roots of complex numbers, ω can be written as $-1/2 + \sqrt{-3}/2$ or as $\cos 120° + i \sin 120°$, depending on the form used for complex numbers, and similarly for ω^2. It may be convenient, given angles θ and ψ that add up to 360°, to view ψ not as $360° - \theta$ but as $-\theta$, keeping in mind that $\cos(360° - \theta) = \cos(-\theta) = \cos\theta$ and $\sin(360° - \theta) = \sin(-\theta) = -\sin\theta$.

Exercise 5.4. Solve the cubic equation

$$y^3 - 3y + 1 = 0.$$

(i) Using Cardano's formula, show that the solutions have the form

$$\sqrt[3]{-\frac{1}{2} + \frac{\sqrt{3}}{2}i} + \sqrt[3]{-\frac{1}{2} - \frac{\sqrt{3}}{2}i}\,.$$

(ii) The numbers inside the cube roots signs are our familiar friends ω and ω^2, the non-real cube roots of 1. Thus, the solution can be rewritten as

$$y = \sqrt[3]{\omega} + \sqrt[3]{\omega^2}.$$

(iii) Find three cube roots of ω and three cube roots of ω^2, expressed in terms of the cosine and sine of suitable angles.

(iv) Form pairs of cube roots in accordance with Exercise 5.1 and add them, taking into account the advice that preceded this exercise.

(v) The three solutions to $y^3 - 3y + 1 = 0$ have the form $2\cos\theta$, for three particular angles θ. What are they?

(vi) We now have precise expressions for the three roots of $y^3 - 3y + 1$, expressed in terms of cosines. To proceed further, use a calculator to determine the values of the cosines, at least approximately. Then substitute the approximate answers into the polynomial $y^3 - 3y + 1$.

As the last part of Exercise 5.4 may suggest, once we turn to calculators, we are settling for approximate answers, and these can be obtained in a variety of ways independent of Cardano's formula. For example, we can solve cubic equations using Newton's method—a standard application of calculus—or turn to a selection of mathematical programs without even worrying about the nature of the underlying algorithms.

The need to approximate already reared its head in our use of the quadratic formula. The finding of square roots of positive real numbers is an algorithmic process, one we carry out until we arrive at as close an approximation to the answer as we wish.

What the quadratic formula does for us is reduce the problem of solving arbitrary quadratic equations to that of calculating square roots. Likewise, Cardano's formula reduces solving cubic equations to the calculation of square and cube roots. It is only at this stage, and because we may be calculating cube roots of complex numbers, that we turn to trigonometry and calculators.

In addition, both the quadratic formula and Cardano's have theoretical value, representing roots of polynomials in terms of their coefficients. We will make use of this in Section 5.3 in obtaining a formula for the discriminant of a cubic polynomial.

5.2 The Resolvent

Given the reduced cubic polynomial $y^3 + py + q$, one can rewrite the terms

$$-\frac{q}{2} \pm \sqrt{\frac{p^3}{27} + \frac{q^2}{4}}$$

whose cube roots we compute in using Cardano's formula as

$$-\frac{q}{2} \pm \frac{\sqrt{q^2 + 4p^3/27}}{2}.$$

These are the roots of the quadratic polynomial $t^2 + qt - p^3/27$, which has $q^2 + 4p^3/27$ as its discriminant. This suggests an alternative approach to the derivation of Cardano's formula, one presented by Euler in *Elements of Algebra* [26, pp. 263–264].

Exercise 5.5. Let b and c be real numbers and let U, V be the roots of the quadratic polynomial $x^2 + bx + c$. From Exercise 2.5, $b = -(U + V)$ and $c = UV$.

(i) Let d be the real cube root of c. Choose cube roots A of U and B of V so that $AB = d$.

(ii) Using the formulas $A^3 = U$, $B^3 = V$, and $AB = d$, expand $(A+B)^3$ and verify that

$$(A + B)^3 = 3d(A + B) - b.$$

(iii) Conclude that Theorem 5.2 holds.

Theorem 5.2. *Let b and c be real numbers and let $\sqrt[3]{c}$ be the real cube root of c. Let U and V be the roots of the quadratic polynomial*

$$x^2 + bx + c.$$

Choose cube roots $\sqrt[3]{U}$ and $\sqrt[3]{V}$ of U and V so that $\sqrt[3]{U} \cdot \sqrt[3]{V} = \sqrt[3]{c}$. Then the sum

$$\sqrt[3]{U} + \sqrt[3]{V}$$

is a root of the cubic polynomial

$$y^3 - 3\sqrt[3]{c}\, y + b.$$

It is now an easy matter to rederive Cardano's formula:

Theorem 5.3 (Cardano again)**.** *Let p and q be real numbers. Let U and V be the roots of the quadratic polynomial*

$$t^2 + qt - \frac{p^3}{27}.$$

Choose cube roots $\sqrt[3]{U}$ and $\sqrt[3]{V}$ of U and V so that $\sqrt[3]{U} \cdot \sqrt[3]{V} = -p/3$. Then the sum

$$\sqrt[3]{U} + \sqrt[3]{V}$$

is a root of the cubic polynomial

$$y^3 + py + q.$$

Exercise 5.6. Verify that Theorem 5.3 is an immediate consequence of Theorem 5.2. Verify also that Theorem 5.3 becomes Cardano's formula in its usual form once we use the quadratic formula to express U and V in terms of p and q.

The polynomial $t^2 + qt - p^3/27$ associated with $y^3 + py + q$ is called its *resolvent quadratic*.

5.3 The Discriminant

For a quadratic polynomial $x^2 + bx + c$ with real or complex roots r_1 and r_2, we saw in Exercise 4.8 and Theorem 4.4 that the squared difference $(r_1 - r_2)^2$ of the roots is expressible in terms of the coefficients:

$$(r_1 - r_2)^2 = b^2 - 4c.$$

This allows us to obtain information on the roots from the coefficients alone: if $b^2 - 4c$ is positive, the roots are real and distinct; if $b^2 - 4c$ is 0, the roots are real and coincide; and if $b^2 - 4c$ is negative, then the roots are a pair of non-real, complex conjugate numbers.

We would like similarly to be able to obtain information about the roots of a cubic polynomial in terms of its coefficients. The discussion that followed Exercise 4.8 showed that the two quantities $b^2 - 4c$ and $(r_1 - r_2)^2$ associated with $x^2 + bx + c$ can both be regarded as its discriminant, but $(r_1 - r_2)^2$ is the more fundamental quantity, the one we chose as the discriminant's definition.

Given a cubic polynomial $x^3 + bx^2 + cx + d$, let's write r_1, r_2, r_3 for its roots, real or complex. In analogy with the quadratic case, we define the *discriminant* of $x^3 + bx^2 + cx + d$ to be

$$(r_1 - r_2)^2(r_1 - r_3)^2(r_2 - r_3)^2$$

and denote it by Δ.

For the special case of a reduced cubic polynomial $x^3 + px + q$, we introduced the quantity $-4p^3 - 27q^2$ in Section 3.4, denoted it by δ, and called it the discriminant. We also saw, in Exercise 3.22 and Theorem 3.6, that the nature of the roots of $x^3 + px + q$ is determined by δ: the polynomial has three distinct real roots if δ is positive, a repeated real root if δ is 0, and one simple real root plus two complex conjugate roots if δ is negative. Let's show for a general cubic polynomial that our newly introduced Δ determines the nature of the roots in the same way.

Exercise 5.7. Let $f(x) = x^3 + bx^2 + cx + d$. We will relate the sign of the discriminant Δ of $f(x)$ to the nature of the roots of $f(x)$.

(i) Suppose $f(x)$ has a multiple root. Show that $\Delta = 0$.

(ii) Suppose that the three roots are real and distinct. Show that $\Delta > 0$.

(iii) Suppose that one root is real and the other two are complex conjugates. Show that $\Delta < 0$. (Hint: Suppose the roots are r, $a + bi$, and $a - bi$, with r, a, and b all real numbers. Using this notation, calculate Δ by calculating the product of the root differences before squaring.)

(iv) Prove the converses: if $\Delta = 0$, then $f(x)$ has a multiple root and all its roots are real; if $\Delta > 0$, then $f(x)$ has three distinct real roots; and if $\Delta < 0$, then $f(x)$ has one real root and two non-real complex conjugate roots. (Hint: These follow from the three statements by the logical argument that we used in Exercise 4.7 to prove the quadratic analogue.)

(v) Conclude that you can determine the nature of the roots of $f(x)$ from the sign of the discriminant of $f(x)$, as described in Theorem 5.4.

Theorem 5.4. *Let $f(x)$ be a degree 3 polynomial with discriminant Δ.*

(i) If Δ is positive, then $f(x)$ has three distinct real roots.

(ii) If $\Delta = 0$, then $f(x)$ has only real roots, one of which occurs with multiplicity at least 2.

(iii) If Δ is negative, then $f(x)$ has three distinct roots. One is real and two form a complex conjugate pair.

We would like to be able to compute the discriminant of a cubic polynomial in terms of its coefficients, so that we can determine the nature of the roots without computing them. Let's first consider a reduced cubic polynomial, one of the form $y^3 + py + q$. We just reviewed that the nature of its roots is determined by the sign of the quantity δ, with $\delta = -4p^3 - 27q^2$. If we can prove that $\delta = \Delta$, then we can put aside any concerns that the two uses of the word "discriminant" conflict. At the same time, we will have an algebraic proof that the quantity $-4p^3 - 27q^2$ determines the nature of the roots. This would contrast with our proof of Theorem 3.6, which depended on calculus.

Exercise 5.8. Suppose p and q are real numbers and consider the polynomial $y^3 + py + q$. We will use the notation and results of Exercise 5.1. Accordingly, we choose A to be one of the three cube roots of $-q/2 + \sqrt{R}$ and B to be the cube root of $-q/2 - \sqrt{R}$ satisfying $AB = -p/3$. With these choices the three roots of $y^3 + py + q$ are

$$r_1 = A + B, \quad r_2 = \omega A + \omega^2 B, \quad r_3 = \omega^2 A + \omega B.$$

(i) Verify that

$$r_1 - r_2 = (1-\omega)(A - \omega^2 B),$$

that

$$r_1 - r_3 = -\omega^2(1-\omega)(A - \omega B),$$

and that

$$r_2 - r_3 = \omega(1-\omega)(A - B).$$

(ii) Verify that

$$(1-\omega)^3 = 3(\omega^2 - \omega) = -3\sqrt{3}i$$

and use Theorem 4.16 to deduce that

$$(r_1 - r_2)(r_1 - r_3)(r_2 - r_3) = -(1-\omega)^3(A^3 - B^3) = 3\sqrt{3}i(A^3 - B^3).$$

(iii) The definitions of A and B yield $A^3 = -q/2 + \sqrt{R}$ and $B^3 = -q/2 - \sqrt{R}$. Deduce that

$$A^3 - B^3 = 2\sqrt{R}$$

and that

$$(r_1 - r_2)(r_1 - r_3)(r_2 - r_3) = 6\sqrt{3}i\sqrt{R}.$$

(iv) Deduce that the discriminant Δ of $y^3 + py + q$ is given by the formula

$$\Delta = -108R.$$

(v) Conclude that Theorem 5.5 holds.

Theorem 5.5. *Given real numbers p and q, the discriminant Δ of the polynomial $y^3 + py + q$ is*

$$-4p^3 - 27q^2.$$

Exercise 5.9. Calculate the discriminant of each cubic polynomial and state whether it has a multiple root, a real root and two non-real roots, or three distinct real roots.

(i) $y^3 - 3y + 2$.

(ii) $y^3 + 5y + 1$.

(iii) $y^3 - 5y + 1$.

It remains to show for a general cubic polynomial $x^3 + bx^2 + cx + d$ that its discriminant Δ can be computed in terms of b, c, and d. This will allow us, as in the quadratic case, to determine the nature of the roots of a cubic from the coefficients alone.

Exercise 5.10. Consider the cubic polynomial $x^3 + bx^2 + cx + d$, with real numbers b, c, and d as coefficients.

(i) From Exercise 3.2 and Theorem 3.3, the change of variable $x = y - (b/3)$ transforms $x^3 + bx^2 + cx + d$ into a polynomial of the form $y^3 + py + q$. Write the roots of $y^3 + py + q$ as r_1, r_2, and r_3. In terms of r_1, r_2, and r_3, what are the roots of $x^3 + bx^2 + cx + d$? (See Exercise 3.2.)

(ii) Observe that although the roots of $x^3 + bx^2 + cx + d$ and $y^3 + px + q$ differ, their differences are the same.

(iii) Deduce that the discriminant of $x^3 + bx^2 + cx + d$ and the discriminant of $y^3 + px + q$ are equal.

(iv) Using the formulas of Theorem 3.3 for p and q in terms of b, c, and d and the formula for the discriminant of $y^3 + px + q$, obtain the formula in Theorem 5.6 for the discriminant of $x^3 + bx^2 + cx + d$ in terms of b, c, and d.

(v) Conclude that the nature of the roots of $x^3 + bx^2 + cx + d$ can be determined in terms of b, c, and d, without explicit knowledge of the roots.

Theorem 5.6. *Given real numbers b, c, and d, the discriminant Δ of $x^3 + bx^2 + cx + d$ is*

$$18bcd - 4b^3d + b^2c^2 - 4c^3 - 27d^2.$$

Exercise 5.11. Calculate the discriminant of each cubic polynomial and state whether it has a multiple root, a real root and two non-real roots, or three distinct real roots.

(i) $x^3 + 2x^2 - 4$.

(ii) $x^3 + 3x^2 - 5x + 1$.

The discriminant separates cubic polynomials into three families according to the nature of their roots, as described in Theorem 5.4: there are three distinct real roots when Δ is positive, there is one real root along with a distinct pair of complex conjugate roots when Δ is negative, and there is a repeated root when Δ is zero. Theorem 5.6 allows us to determine which occurs for a cubic polynomial from its coefficients alone. If $\Delta = 0$, we can use the coefficients to do a little more.

Theorem 5.7. *The polynomial $x^3 + bx^2 + cx + d$ has one real root, of multiplicity 3, if $\Delta = 0$ and $b^2 - 3c = 0$. It has two real roots, of multiplicities 1 and 2, if $\Delta = 0$ and $b^2 - 3c \neq 0$.*

Exercise 5.12. Prove Theorem 5.7.

(i) Check that the only reduced cubic polynomial with a root of multiplicity 3 is y^3. (Hint: Given a real number r, what is the coefficient of y^2 in $(y - r)^3$?)

(ii) Deduce that $y^3 + py + q$ has a root of multiplicity 3 if $\Delta = 0$ and $p = 0$, but roots of multiplicities 1 and 2 if $\Delta = 0$ and $p \neq 0$.

(iii) Observe that $x^3 + bx^2 + cx + d$ has a root of multiplicity 3 precisely when the reduced cubic polynomial associated with it by the change of variable of Theorem 3.3 does.

(iv) Show that the theorem follows.

The sign of a cubic's discriminant determines whether or not we will have to calculate cube roots of non-real complex numbers when using Cardano's formula. We saw this at the conclusion of Section 3.4 in terms of δ, the discriminant we obtained using calculus. Having now developed the theory of the discriminant algebraically, we can restate the situation in terms of Δ.

Exercise 5.13. Consider the reduced cubic polynomial $y^3 + py + q$. As usual, let $R = p^3/27 + q^2/4$.

(i) Using Theorem 5.5, observe that Δ and R have opposite signs.

(ii) Review the relation of the sign of Δ to the nature of the roots and rephrase it as a relation between the sign of R and the nature of the roots.

(iii) Deduce that if $y^3 + py + q$ has one real root and two non-real roots, then Cardano's formula expresses the real root as the sum of cube roots of real numbers.

(iv) In contrast, deduce that if y^3+py+q has three distinct real roots, then Cardano's formula for the roots expresses all three of them as sums of cube roots of non-real complex numbers.

The appearance of cube roots of non-real complex numbers in Cardano's formula when Δ is positive can be made explicit by rewriting Cardano's formula with Δ in place of R:

$$y = \sqrt[3]{-\frac{q}{2} + \frac{\sqrt{-\Delta}}{6\sqrt{3}}} + \sqrt[3]{-\frac{q}{2} - \frac{\sqrt{-\Delta}}{6\sqrt{3}}}.$$

Our examples of polynomials for which such cube roots arose weren't anomalies. They were inevitable.

5.4 Cardano's Formula Refined

The nature of a reduced cubic polynomial's roots depends on the sign of its discriminant. For each of the three possible cases—zero, negative, or positive—we can obtain a refined version of Cardano's formula. Let's start with the case of zero discriminant.

Exercise 5.14. Let $y^3 + py + q$ be a reduced cubic polynomial with discriminant equal to 0; that is, $-4p^3 - 27q^2 = 0$. The case $p = q = 0$ is not interesting, so let's assume p and q are both non-zero.

(i) Show that p must be negative.

(ii) Solve the equation $-4p^3 - 27q^2 = 0$ for q in terms of p. Show that

$$q = \pm\frac{2p}{3}\sqrt{-\frac{p}{3}}.$$

(iii) To simplify notation, let's introduce a new constant a, with $a = \sqrt{-p/3}$; that is, a is the positive square root of $-p/3$. Check that $p = -3a^2$ and $q = \pm 2a^3$. The choice of sign in the expression for q in terms of a is determined by the sign of q: if q is positive, then $q = 2a^3$; if q is negative, then $q = -2a^3$.

(iv) Assume that q is positive. Rewrite the cubic polynomial $y^3 + py + q$ as

$$y^3 - 3a^2y + 2a^3.$$

Verify that $y = a$ and $y = -2a$ are roots. Show that we have the factorization

$$y^3 - 3a^2y + 2a^3 = (y-a)^2(y+2a).$$

(v) Assume that q is negative. Rewrite $y^3 + py + q$ as

$$y^3 - 3a^2y - 2a^3.$$

Verify that its roots are $-a$ and $2a$, and that we have the factorization

$$y^3 - 3a^2y - 2a^3 = (y+a)^2(y-2a).$$

(vi) Deduce Theorem 5.8.

Theorem 5.8. *Let $y^3 + py + q$ be a reduced cubic polynomial with discriminant equal to* 0.

(i) If $q = 0$, then $p = 0$ and 0 is a root of multiplicity 3.

(ii) If $q > 0$, then $\sqrt{-p/3}$ is a root of multiplicity 2 and $-2\sqrt{-p/3}$ is a root of multiplicity 1.

(iii) If $q < 0$, then $-\sqrt{-p/3}$ is a root of multiplicity 2 and $2\sqrt{-p/3}$ is a root of multiplicity 1.

Let's next consider the case of negative discriminant. We saw in Exercise 4.27 that a real number c has one real cube root, say d, and two non-real cube roots, ωd and $\omega^2 d$. This allows us to be more explicit about our choices of A and B in the version of Cardano's formula stated as Theorem 5.1.

Exercise 5.15. Let $y^3 + py + q$ be a reduced cubic polynomial with negative discriminant; that is, $-4p^3 - 27q^2 < 0$. Let R as usual equal $p^3/27 + q^2/4$, so $\Delta = -108R$ and $R > 0$. In particular, the quantities $-q/2 + \sqrt{R}$ and $-q/2 - \sqrt{R}$ are real. Let A be the real cube root of $-q/2 + \sqrt{R}$ and let B be the real cube root of $-q/2 - \sqrt{R}$.

(i) Verify that $AB = -p/3$. (This can be done directly. Or, from Exercise 5.1, the product of A with one of B, ωB, and $\omega^2 B$ is $-p/3$. Consider which product is a real number.)

(ii) Deduce in the case of negative discriminant that Theorem 5.1 takes on the special form of Theorem 5.9.

Theorem 5.9. *Let y^3+py+q be a reduced cubic polynomial with negative discriminant. Let A be the real cube root of*

$$-\frac{q}{2}+\sqrt{\frac{p^3}{27}+\frac{q^2}{4}},$$

let B be the real cube root of

$$-\frac{q}{2}-\sqrt{\frac{p^3}{27}+\frac{q^2}{4}},$$

and let $\omega=-1/2+\sqrt{-3}/2$. Then y^3+py+q has one real root, $A+B$, and two non-real roots, $\omega A+\omega^2 B$ and $\omega^2 A+\omega B$.

In using Cardano's formula to find the roots of a cubic polynomial y^3+py+q with positive discriminant, we can cut the work of cube root calculations in half once we recognize that the two expressions whose cube roots we need to calculate are complex conjugates of each other.

Exercise 5.16. Let y^3+py+q be a reduced cubic polynomial with positive discriminant; that is, $-4p^3-27q^2>0$. Let $R=p^3/27+q^2/4$, so that $R<0$.

(i) Since R is negative, $-R$ has two real square roots, $\sqrt{-R}$ and $-\sqrt{-R}$. Observe that $-q/2+\sqrt{-R}\,i$ and $-q/2-\sqrt{-R}\,i$ are complex conjugates.

(ii) If s is a complex number and r is one of its cube roots, then Exercise 4.1 yields that $\overline{r}$ is a cube root of $\overline{s}$. Suppose A is a cube root of $-q/2+\sqrt{-R}\,i$. Deduce that $\overline{A}$ is a cube root of $-q/2-\sqrt{-R}\,i$.

(iii) Verify that $A\overline{A}=-p/3$. (Use the fact that the product of A with one of $\overline{A}$, $\omega\overline{A}$, and $\omega^2\overline{A}$ is $-p/3$. Consider which product is a real number.)

(iv) Use Exercise 4.4 to verify that ωA and $\omega^2\overline{A}$ are complex conjugates and that $\omega^2 A$ and $\omega\overline{A}$ are complex conjugates.

(v) Deduce that Theorem 5.1 takes on the special form of Theorem 5.10.

Theorem 5.10. *Let y^3+py+q be a reduced cubic polynomial with positive discriminant. Let A be a cube root of*

$$-\frac{q}{2}+\sqrt{\frac{p^3}{27}+\frac{q^2}{4}}$$

and let ω be the cube root $-1/2+\sqrt{-3}/2$ of 1. Then the three roots of y^3+py+q are $A+\overline{A}$, $\omega A+\overline{\omega A}$, and $\omega^2 A+\overline{\omega^2 A}$.

Let us return to the cubic $y^3 - 7y + 6$ and find its roots using Theorem 5.10.

Exercise 5.17. Use Cardano's formula, as refined in Theorem 5.10, to solve the cubic equation

$$y^3 - 7y + 6 = 0.$$

(i) One cube root of $-3 + 10\sqrt{-3}/9$ was determined in Exercise 3.13 or Exercise 4.28. Add this cube root to its conjugate to obtain one solution to $y^3 - 7y + 6 = 0$.

(ii) Using ω and ω^2, write expressions for the other two cube roots of $-3 + 10\sqrt{-3}/9$. Add each to its conjugate to obtain the other two solutions.

We close this section with two exercises that provide an opportunity to tie together the ideas we have learned so far.

Exercise 5.18. Solve the equation

$$x^3 + 6x^2 + 3x + 18 = 0.$$

The solution should be presented in essay form, with explanations of what is being done at each step and why it is being done. The solution should not be merely lines of calculation. There should be English prose. The following steps should be included.

(i) Change variables to obtain a presumably simpler cubic equation to solve.

(ii) State Cardano's formula clearly in general and then explain how to apply it.

(iii) Use Cardano's formula to obtain a real solution.

(iv) Obtain the other two solutions in two ways. First use factorization and the quadratic formula. Then obtain the other two solutions using cube roots of 1 and Cardano's formula again.

(v) Obtain three solutions to the original equation.

Exercise 5.19. Solve the equation

$$x^3 - 3x^2 - 10x + 24 = 0.$$

Again the solution should be presented as an essay. The points listed in Exercise 5.18 apply to this equation as well, but there are some differences in the details:

(i) This time it may be necessary to calculate cube roots of non-real complex numbers, which can be done using trigonometry, along with a calculator.

(ii) Calculator solutions are only approximate. Point this out. Even though the solution is approximate, it may in fact be correct. Check to see if it is.

5.5 The Irreducible Case

Reduced cubic polynomials with positive discriminant are precisely those with three distinct real roots, as we saw in Theorem 5.4. Thus, they factor as products of three distinct linear polynomials and are said to be *completely reducible* in modern terminology. Yet, they are also the polynomials whose roots Cardano's formula expresses as sums of cube roots of non-real complex numbers. Cardano called this the *irreducible case*. To understand the sense in which "irreducible" applies, let's continue an analysis that we began in Exercises 4.13 and 4.14.

Exercise 5.20. Fix real numbers a and b with $b \neq 0$. We wish to calculate cube roots of $a + bi$. This amounts to solving the equation $(x + yi)^3 = a + bi$ for real values of x and y,

(i) Review Exercises 4.13 and 4.14, where we found that the equation $(x + yi)^3 = a + bi$ yields two equations for x and y:

$$x^3 - 3xy^2 = a; \quad 3x^2y - y^3 = b.$$

(ii) The substitution $x = sy$ led to a single equation in s,

$$s^3 - 3\frac{a}{b}s^2 - 3s + \frac{a}{b} = 0.$$

The equation is a cubic in s with coefficients expressed in terms of a/b.

(iii) Using Cardano's formula to solve a cubic equation of "irreducible" type requires us to calculate a cube root of a non-real complex number. We now find that calculating such a cube root leads to a new cubic equation. Have we made progress? Let's examine the new cubic.

(iv) Introduce another variable t, with $s = t + (a/b)$. Substitute $t + (a/b)$ for s in the equation for s to obtain a reduced cubic equation for t,

$$t^3 - 3kt - 2\frac{a}{b}k = 0,$$

with $k = 1 + (a^2/b^2)$.

(v) We can try to solve it using Cardano's formula. Calculate R, or Δ for this new reduced cubic. Verify that R has the form $-k^2$. This means the new cubic, in t, is also of irreducible type, so that solving it will require another calculation of the cube root of a non-real complex number.

(vi) Show that in fact using Cardano's formula to solve the cubic in t requires us to compute the cube root of

$$\frac{a}{b}k + ki.$$

(vii) Write this as

$$\frac{k}{b} \cdot (a + bi),$$

a real number times $a + bi$.

(viii) Conclude that in order to use Cardano's formula to find roots of the reduced cubic equation in t, we must be able to compute the cube root of none other than $a + bi$, the complex number whose cube root we were trying to calculate in the first place!

The calculation in the last exercise shows that we have come full circle. To solve an "irreducible" cubic equation, we need to calculate the cube root of a non-real complex number $a + bi$. To compute it, we are led to solve a different cubic equation. Passing from this different cubic to its associated reduced cubic equation and applying Cardano's formula, we find that we must compute the same cube root with which we began. It is in this sense that the problem is irreducible.

We need not despair. We must accept that non-algebraic techniques are needed. For example, we can use trigonometry to compute cube roots, as worked out in Exercise 4.27.

5.6 Viete's Formula

We closed Section 5.5 with the recognition that we must leave the realm of algebra for trigonometry to find the roots of reduced cubic polynomials in the irreducible case. We previously observed (Exercise 5.16 and Theorem 5.10) in this case that the two summands in Cardano's formula are complex conjugates of each other. The sum of a complex number and its conjugate is twice the real part of the two numbers: $(a + bi) + (a - bi) = 2a$. Thus, given a cubic polynomial $y^3 + py + q$ with positive discriminant, if we can determine the real parts of the cube roots of $-q/2 + \sqrt{R}$ in terms of p and q, we will have a new formula for the cubic's roots. We will obtain such a formula, using the standard trigonometric functions and inverse trigonometric functions.

From the discussion before Exercise 4.16, the inverse cosine function $\arccos x$ is defined only for real numbers in the interval $[-1, 1]$. As x increases from -1 to 1, the value $\arccos x$ decreases from π to 0, taking each value once. The following inequality ensures that the inverse cosine is defined on expressions that will arise in Exercise 5.22.

Exercise 5.21. Suppose p and q are real numbers. Assume $p < 0$ and let $\sqrt{-p/3}$ be the positive square root of $-p/3$. Show that

$$-1 < \frac{3q}{2p\sqrt{-p/3}} < 1$$

precisely when $-4p^3 - 27q^2 > 0$. (Hint: Use the fact, for a real number r, that $-1 < r < 1$ precisely when $r^2 < 1$.)

Exercise 5.22. Suppose p and q are real numbers satisfying $-4p^3-27q^2 > 0$. Let $R = p^3/27 + q^2/4$ (which is negative) and take $\sqrt{-R}$ to be the positive square root of $-R$.

(i) Observe that p must be negative. Write $\sqrt{-p/3}$ for the the positive square root of $-p/3$.

(ii) By Exercise 4.17, we can write

$$-\frac{q}{2} + \sqrt{-R}i$$

in the form

$$r \cos\theta + (r \sin\theta)i$$

for real numbers r and θ, with $r > 0$. Using Exercise 4.16, show that

$$r = \sqrt{-\frac{p^3}{27}} = -\frac{p}{3}\sqrt{-\frac{p}{3}},$$

and that one choice of θ is

$$\theta = \arccos\left(-\frac{q}{2r}\right) = \arccos\left(\frac{3q}{2p\sqrt{-p/3}}\right).$$

(By Exercise 5.21, arccos is defined here.)

(iii) Let's simplify our notation by introducing new constants a and b, with $a = \sqrt{-p/3}$ and $b = 3q/p$. Then $r = a^3$ and $\theta = \arccos(b/2a)$.

(iv) From Exercise 4.27, one cube root of

$$r \cos\theta + ri \sin\theta$$

is

$$\sqrt[3]{r}\,\cos\frac{\theta}{3} + \sqrt[3]{r}\,i \sin\frac{\theta}{3}.$$

Deduce, using our shorthand notation, that

$$a \cos\left(\frac{1}{3}\arccos\left(\frac{b}{2a}\right)\right) + ai \sin\left(\frac{1}{3}\arccos\left(\frac{b}{2a}\right)\right)$$

is a cube root of $-q/2 + \sqrt{-R}i$.

(v) By Theorem 5.10, one root of $y^3 + py + q$ is given by adding the complex number displayed in (iv) to its conjugate. Adding a complex number and its complex conjugate yields twice the real part of the complex number. Conclude that

$$2a \cos\left(\frac{1}{3}\arccos\left(\frac{b}{2a}\right)\right)$$

is a root of $y^3 + py + q$.

The formula for a root of the reduced cubic $y^3 + py + q$ in Exercise 5.22 is due to François Viète, whose work we will discuss further in Section 5.8. Let's record it as a theorem, replacing a and b with the expressions in p and q that they represent.

Theorem 5.11 (Viète). *Suppose p and q are real numbers. Let $y^3 + py + q$ be a cubic polynomial with positive discriminant. Then one of the roots of $y^3 + py + q$ is given by the formula*

$$2\sqrt{-\frac{p}{3}}\cos\left(\frac{1}{3}\arccos\left(\frac{3q}{2p\sqrt{-p/3}}\right)\right).$$

From Exercise 4.27, we can express the other two roots of $y^3 + py + q$ by similar expressions, by adding $2\pi/3$ and $4\pi/3$ to the argument of the cosine.

Viète's formula allows us to avoid calculating cube roots of non-real complex numbers in finding roots of cubic polynomials of irreducible type. This is reminiscent of the situation we discussed in Section 4.8, where we examined how to calculate nth roots of real and complex numbers using the exponential and logarithm functions or using the cosine and inverse cosine functions. We concluded that discussion with the observation that in both cases the calculation involved using a familiar function f—the exponential or cosine function—in the same way. We calculate the value of f^{-1} for some constant c, divide the resulting number by n, and then apply f to the result or a minor variant. For example, the nth root of a positive real number r is

$$e^{\log(r)/n}.$$

Viète's formula fits into this theme, with the cosine function as f. We calculate an inverse cosine, divide by 3, and calculate a cosine, thereby replacing a cube root calculation (of a complex number!) with division by 3. Let's try it.

Exercise 5.23. Verify that $y^3 - 7y + 6$ has positive discriminant. Then use Viète's formula and a calculator to solve

$$y^3 - 7y + 6 = 0.$$

Exercise 5.24. Verify that $y^3 - 3y + 1$ has positive discriminant. Then use Viète's formula to express one of the solutions to the equation

$$y^3 - 3y + 1 = 0$$

in terms of cosines. We get the same answer as in Exercise 5.4 using Cardano's and de Moivre's formulas.

Viète's formula applies also to a cubic $y^3 + py + q$ with zero discriminant. We can rework Exercise 5.21 (still assuming $p < 0$) to find that $-4p^3 - 27q^2 = 0$ precisely when $3q/2p\sqrt{-p/3}$ equals 1 or -1. We obtain -1 when q is positive and 1 when q is negative. The inverse cosine of -1 is π and of 1 is 0. Thus, when $q > 0$, Viète's formula yields $y = \sqrt{-p/3}$, while when $q < 0$, the formula yields $y = 2\sqrt{-p/3}$. This proves again part of Theorem 5.8.

Although Viète's formula allows us to bypass the calculation of cube roots of non-real complex numbers, such cube roots did appear in its derivation. Once we know—or guess—that a formula for y exists expressing it as twice a positive multiple of a cosine, we may wish to derive it directly, without reference to Cardano's formula or complex numbers. This can be done, using the cosine triple angle formula:

Theorem 5.12. *Let θ be a real number. Then*

$$\cos 3\theta = 4\cos^3\theta - 3\cos\theta.$$

Exercise 5.25. Prove Theorem 5.12.

(i) Using the angle sum formulas of Theorem 4.9, write formulas for $\sin 2\theta$ and $\cos 2\theta$ in terms of $\sin\theta$ and $\cos\theta$.

(ii) Using Theorem 4.9 again, write $\cos 3\theta$ in terms of sines and cosines of θ and 2θ.

(iii) Obtain the identity

$$\cos 3\theta = \cos^3\theta - 3\sin^2\theta\,\cos\theta.$$

(iv) Use

$$\cos^2\theta + \sin^2\theta = 1$$

and the equality in (iii) to obtain the desired identity.

We can now use Theorem 5.12 to derive Viète's formula independently of Cardano's formula.

Exercise 5.26. Let $y^3 + py + q$ be a cubic polynomial with positive discriminant. Recall that p must be negative. We wish to find a solution in the form $y = 2a\cos\theta$, where a and θ are real numbers expressed in terms of p and q and a is positive. (We have been using a to represent $\sqrt{-p/3}$, but for now we assume that a is unknown).

(i) Substitute $2a\cos\theta$ for y in the equation $y^3 + py + q = 0$ to obtain

$$8a^3\cos^3\theta + 2ap\cos\theta + q = 0.$$

If we can find values of a and θ, in terms of p and q, for which this holds, then we will have the desired formula.

(ii) Use the triple angle formula and collect terms to rewrite the equation as

$$(6a^3 + 2ap)\cos\theta + 2a^3\cos 3\theta + q = 0.$$

It will suffice to find values of a and θ for which $6a^3 + 2ap = 0$ and $2a^3\cos 3\theta + q = 0$.

(iii) We have assumed that a is positive, so that $a \neq 0$. Using this, solve $6a^3 + 2ap = 0$ for a and show that

$$a = \sqrt{-p/3}.$$

(iv) Substitute $\sqrt{-p/3}$ for a in the equation

$$2a^3\cos 3\theta + q = 0,$$

solve for θ, and show that

$$\theta = \frac{1}{3}\arccos\left(\frac{3q}{2p\sqrt{-p/3}}\right).$$

(v) Rewrite $2a\cos\theta$ as an expression in p and q using the expressions we have just obtained for a and θ.

We have succeeded in deriving Viète's formula directly, without introducing complex numbers.

5.7 The Signs of the Real Roots

We have seen how to compute the discriminant Δ of $x^3 + bx^2 + cx + d$ in terms of the coefficients b, c, and d. We have also seen how to use Δ to decide whether x^3+bx^2+cx+d has a repeated root, three distinct real roots, or one real root and two complex conjugate roots. Thus this information can be determined using only the coefficients.

It is possible to obtain more information from the coefficients, such as how many of the real roots are positive and how many are negative. We will close our treatment of cubic polynomials by working this out, because it is interesting in its own right and because we will need the result in our study of quartic polynomials. We begin by obtaining analogous results for linear and quadratic polynomials.

Exercise 5.27. Let c be a real number. Show that the only root of the linear polynomial $x + c$ is $-c$. Deduce that the sign of the root can be determined in terms of the sign of the constant coefficient. (Yes, this is trivial, but it is the first in a sequence of results.)

Exercise 5.28. Let b and c be real numbers and let r_1 and r_2 be the roots of the quadratic polynomial x^2+bx+c. Since we are interested in the case that the roots are real, we assume that $b^2-4c \geq 0$.

(i) If $c=0$, the roots are 0 and $-b$, so their sign is determined by the sign of the coefficient b.

(ii) Assume $c \neq 0$. From Exercise 2.5, $c = r_1 r_2$ and $b = -(r_1+r_2)$. Show that if r_1 and r_2 have the same sign, then $c > 0$, and if they have opposite sign, then $c < 0$.

(iii) Assume that c is positive. Show that r_1 and r_2 are both positive precisely when $b < 0$ and both negative precisely when $b > 0$.

(iv) Deduce Theorem 5.13.

Theorem 5.13. *Let b and c be real numbers satisfying $b^2-4c \geq 0$ and $c \neq 0$.*

(i) The polynomial x^2+bx+c has two real roots, both non-zero.

(ii) If $c > 0$ and $b < 0$, then both roots are positive.

(iii) If $c < 0$, then one root is positive and one is negative.

(iv) If $c > 0$ and $b > 0$, then both roots are negative.

To handle cubic polynomials, we need a cubic analogue of Exercise 2.5.

Theorem 5.14. *Let x^3+bx^2+cx+d be a cubic polynomial with real coefficients b, c, and d, and suppose its three roots are r_1, r_2, and r_3. Then*

$$\begin{aligned} b &= -r_1 - r_2 - r_3, \\ c &= +r_1 r_2 + r_1 r_3 + r_2 r_3, \\ d &= -r_1 r_2 r_3. \end{aligned}$$

Exercise 5.29. Prove Theorem 5.14. (Hint: Use the factorization of x^3+bx^2+cx+d as $(x-r_1)(x-r_2)(x-r_3)$.)

A cubic polynomial of the form x^3+bx^2+cx factors as the product of x and x^2+bx+c, so that one of its roots is 0 and the others are roots of x^2+bx+c. Thus, we can determine the signs of its roots using Theorem 5.13. We will exclude this case.

Let's consider cubic polynomials that have non-real roots.

Theorem 5.15. *Let $x^3 + bx^2 + cx + d$ be a cubic polynomial with real coefficients b, c, and d whose discriminant is negative. Assume further that $d \neq 0$. Then the real root of $x^3 + bx^2 + cx + d$ is positive if d is negative and negative if d is positive.*

Exercise 5.30. Prove Theorem 5.15. (Hint: Use Theorem 5.14 and the fact that the product of a non-zero complex number and its conjugate is a positive real number.)

We now turn to cubic polynomials whose roots are all real.

Exercise 5.31. Assume that x^3+bx^2+cx+d is a cubic polynomial with real coefficients b, c, and d. Assume that $d \neq 0$ and that the discriminant Δ of x^3+bx^2+cx+d is non-negative, so that the roots of x^3+bx^2+cx+d are all real (whatever their multiplicities).

(i) Show that if b, c, and d are positive, then all the roots are negative. (Hint: If $x > 0$, what is the sign of $x^3 + bx^2 + cx + d$?)

(ii) Conversely, show that if all the roots are negative, then b, c, and d are positive. (Hint: Examine the expressions for b, c, and d in terms of the roots given in Theorem 5.14.)

(iii) Show that if b is negative, c is positive, and d is negative, then all the roots are positive. (Hint: If $x < 0$, what is the sign of x^3+bx^2+cx+d?)

(iv) Conversely, show that if all the roots are positive, then b is negative, c is positive, and d is negative.

(v) Show that $d > 0$ if there are 0 or 2 positive roots and $d < 0$ if there are 1 or 3 positive roots.

(vi) Deduce Theorem 5.16.

Theorem 5.16. *Let $x^3 + bx^2 + cx + d$ be a cubic polynomial with real coefficients b, c, and d whose discriminant is non-negative. Assume further that $d \neq 0$.*

(i) If b, c, and d are positive, then all the roots are negative.

(ii) If b is negative, c is positive, and d is negative, then all the roots are positive.

(iii) If neither condition holds, then there are both positive and negative roots. In this case, if $d > 0$, then there are two positive roots; if $d < 0$, then there is one positive root.

In the next exercise, we will apply what we've learned.

Exercise 5.32. For the cubic polynomials in (i)-(iv), use Theorems 5.15 and 5.16 to determine from the coefficients how many real roots there are and how many of them are positive or negative.

(i) $x^3 - 6x^2 + 11x - 6$.

(ii) $x^3 - 5x^2 + 9x - 5$.

(iii) $x^3 + 6x^2 + 3x + 18$.

(iv) $x^3 - 3x^2 - 10x + 24$.

Let's extract one small piece of Theorems 5.15 and 5.16 that will play an essential role in our study of quartic polynomials. (See Exercise 6.8.)

Theorem 5.17. *Let $f(x)$ be a cubic polynomial with non-zero constant coefficient d.*

(i) If $d > 0$, then $f(x)$ has at least one negative real root.

(ii) If $d < 0$, then $f(x)$ has at least one positive real root.

Exercise 5.33. Verify that Theorem 5.17 follows from Theorems 5.15 and 5.16. It is valid regardless of whether $f(x)$ has one or three real roots.

5.8 History

We concluded Section 5.5 with the observation that we need not despair when faced with a cubic equation to solve in the irreducible case. How did Cardano react? Let's find out, as we continue the historical account of cubic equations that we began in Section 3.5.

The cubic equations whose solution Cardano learned from Tartaglia have the special form

$$x^3 + px = q$$

with p and q positive. Since p is positive, the discriminant Δ is negative and the formula of del Ferro and Tartaglia involves cube roots of real numbers. Cardano, however, extended Tartaglia's solution to cover cubic equations of the additional forms

$$x^3 = px + q$$

and

$$x^3 + q = px.$$

Again, the coefficients are taken to be positive. But then, rewriting them in the form $x^3 + Px + Q = 0$ by moving the appropriate terms to the left side, we see that P is negative, so the discriminant $-4P^3 - 27Q^2$ may be positive. Thus Cardano opened the door to the irreducible case, with its square roots of negative numbers and cube roots of non-real numbers.

Cardano was aware of the new difficulty. However, uncertain how to handle it, he omitted examples of cubic equations of this type in *Ars Magna* [12]. Thus, the answer to our opening question is that he suppressed the irreducible case.

In an interesting passage later in *Ars Magna*, Cardano does make some calculations with square roots of negative numbers. We know from Section 2.1 that solving a quadratic equation $x^2 - bx + c = 0$ (the minus sign in front of the b is intentional) is equivalent to finding two numbers whose sum is b and whose product is c, and that special cases of this problem can be found on Babylonian clay tablets from over three thousand years ago. In Chapter 37, titled "On the Rule for Postulating a Negative," Cardano poses and solves just such a problem [12, p. 219]:

> If it should be said, Divide 10 into two parts the product of which is 30 or 40, it is clear that this case is impossible. Nevertheless, we will work thus: We divide 10 into two equal parts, making each 5. These we square, making 25. Subtract 40 if you will, from the 25 thus produced … leaving a remainder of -15, the square root of which added to or subtracted from 5 gives parts the product of which is 40. These will be $5 + \sqrt{-15}$ and $5 - \sqrt{-15}$.

By competing the square, Cardano has found two numbers that have sum 10 and product 40. Equivalently, he has solved the quadratic equation

$$x^2 - 10x + 40 = 0,$$

despite its discriminant being negative. Next Cardano writes [12, pp. 219–220],

> Putting aside the mental tortures involved, multiply $5 + \sqrt{-15}$ by $5 - \sqrt{-15}$, making $25 - (-15)$, which is $+15$. Hence this product is 40. … This truly is sophisticated … . So progresses arithmetic subtlety the end of which, as is said, is as refined as it is useless.

Recognizing that the problem makes no physical sense, Cardano is content to label the result useless.

Complex numbers enter into the solving of quadratic equations precisely when the equations have no real solutions. Though sophisticated, the new numbers aren't needed. What's different about cubic equations is that the new numbers appear precisely when the equations have only real (and distinct) solutions. This was a puzzle Cardano was not prepared to address.

In contrast, Cardano's near-contemporary Rafael Bombelli (1526–1572) did not shy away from the irreducible case. In the influential three-volume work *L'Algebra* [11], printed in 1572 and again in 1579, Bombelli set out to present the results of *Ars Magna* in a manner more accessible to beginners.

Bombelli gives as one example the irreducible cubic equation

$$x^3 = 15x + 4.$$

(See the discussion in [66, pp. 60–61].) Using Cardano's formula, Bombelli writes the solution

$$\sqrt[3]{2 + \sqrt{-121}} + \sqrt[3]{2 - \sqrt{-121}}.$$

He describes the imaginary square roots, following Cardano, as "sophistic," but he also notes that the cubic equation can be solved, since after all $x = 4$ is a solution. This motivates finding a way to make the formula work, and Bombelli proceeds much as we did in Exercise 3.13 when we used Cardano's formula to solve the cubic equation $y^3 - 7y + 6 = 0$.

Let's write $a + \sqrt{-b}$ as a possible cube root of $2 + \sqrt{-121}$ and $a - \sqrt{-b}$ as the corresponding cube root of $2 - \sqrt{-121}$. By setting the cube of $a + \sqrt{-b}$ equal to $2 + \sqrt{-121}$ and collecting terms, we get

$$a^3 - 3ab = 2.$$

By setting the product of $a + \sqrt{-b}$ and $a - \sqrt{-b}$ equal to the product of $\sqrt[3]{2 + \sqrt{-121}}$ and $\sqrt[3]{2 - \sqrt{-121}}$, we get

$$a^2 + b = 5.$$

Bombelli found the solution $a = 2$ and $b = 1$ to this pair of equations. Adding $2 + \sqrt{-1}$ and $2 - \sqrt{-1}$, he obtained $x = 4$ as the solution to the cubic equation, prompting him to comment, "At first, the thing seemed to me to be based more on sophism than on truth, but I searched until I found the proof." [66, p. 61]

Bombelli also gave rules for the manipulation of these new numbers, including the rule that

$$\sqrt{-1} \cdot \sqrt{-1} = -1,$$

and provided examples of calculations involving them. He didn't use our notation, but the essential ideas were all there.

We have focused on Italian mathematicians, but as the sixteenth-century neared its end, the work of the French mathematician François Viète came to the fore. Some of his writings are available in a 1983 translation by T. Richard Witmer under the title *The Analytic Art* [68]. In *In Artem Analyticem Isagoge*, which appeared in 1591 and is translated by Witmer as *Introduction to the Analytic Art*, Viète introduced the idea of using letters for known and unknown quantities, differentiating between the two by using vowels for unknown quantities and consonants for the known quantities. This allowed him to treat algebraic problems in generality rather than describing general methods implicitly through specific examples.

Another of Viète's contributions, which we studied in Section 5.6, was the development of a trigonometric approach to cubic equations in the irreducible case that allows us to avoid complex numbers altogether. With A as his variable and B and D as constants, he studies ([68, p. 174]) the equation

$$A^3 - 3B^2A = B^2D$$

under the assumption that "B is greater than half of D." Taking B and D to be positive, we can check that the condition $B > D/2$ is the requirement that the given cubic's discriminant is positive. He then describes how to obtain a solution geometrically in terms of two right triangles, one having an acute angle three times that of the second. Next, Viète illustrates the method by solving the cubic equations

$$x^3 - 300x = 432$$

and

$$x^3 - 300x = -432.$$

In [48, pp. 203–205], R.W.D. Nickalls relates this passage of Viète's to our more familiar trigonometric view, explaining that "Viète's approach stems from his familiarity with the then equivalent of the trigonometric triple-angle identities, since he himself had established formulae for chords of multiple arcs in terms of chords of simple arcs, and hence he was aware that solving a cubic with three real roots was analogous to trisecting an angle."

Before moving on, let us recognize the greater significance of Viète's work, as he was the first to write down cubic equations (in the irreducible case) in general form and solve them.

Another French mathematician, Albert Girard (1595–1632), who would spend most of his life in the Netherlands as a religious refugee, published *Invention Nouvelle en l'Algèbre* in 1629 [32]. It's a short work, just 64 pages, with a central section on algebra that is full of new ideas. We will discuss Girard's contributions further in Sections 7.2 and 7.4. For now, we will be content with a few words about his account of cubic equations.

Girard doesn't use letters for constants or variables, yet his text is easy to understand. For example, he divides cubic equations into a variety of cases, one of which he describes as the equations with "1(3) equal to (1) + (0)." The parenthetical numbers should be interpreted as powers of x, so that this refers to equations in which a cube equals a degree-one term plus a constant, or $x^3 = px + q$. For this case, Girard provides a "rule for solving the equation 1(3) equal to (1) + (0) when the cube of a third of the number of (1) is larger than the square of half of (0) with the aid of a table of sines." This is, of course, the irreducible case. Without a general notation, Girard must describe the rule by way of example. The example he uses is "1(3) equal to 13(1) + 12," which we would write as $x^3 = 13x + 12$. He displays a series of calculations that leads to the solution $x = 4$, using what is essentially Viète's approach.

We mentioned John Wallis in passing in Section 4.9 with regard to his early effort to provide a geometric description of complex numbers. This is contained in his wonderfully titled 1685 book, *A Treatise of Algebra, both Historical and Practical: Shewing the original, progress, and advancement thereof, from time to time, and by what steps it hath attained to the heighth at which now it is; with some additional treatises*, which would have great influence. J.F. Scott's *The Mathematical Work of John Wallis*, written in 1938, is a valuable guide to Wallis's mathematical work, with a chapter devoted to the *Treatise of Algebra* [58, pp. 133–165]. Scott's concluding appraisal [58, p. 165] is that Wallis's book "constituted a reservoir from which contemporary and later algebraists drew much inspiration. It may be mentioned that this treatise did a great deal towards popularizing the notation which was now rapidly becoming current in Europe."

Of particular significance is Wallis's unreserved adoption of complex numbers as solutions to algebraic equations. Scott writes [58, pp. 156–157], "In his quadratic equations he discusses every type, and the rules he evolves for determining the nature of the roots by a mere inspection of the equation would not be out of place in a modern text-book. He was quite at home with imaginaries, and he knew that such roots always occurred in pairs. Moreover he would not allow the use of the word *Impossible* as applied to an equation with imaginary roots. An equation, for example, such as

$$-aa + 8a = 25,$$

of which the roots are $4 + \sqrt{16 - 25}$ and $4 - \sqrt{16 - 25}$, 'imaginaries' —

had hitherto been styled 'Impossible'. Yet, avers Wallis" (and now we turn to Wallis's own writings [69, p. 174]):

> Imaginary quantities, when they occur, have been taught to imply an impossible case; and algebraists have been wont so to teach. Yet is not this so to be understood ... . For it was before thought (and so delivered by diverse algebraists) that whenever (in pursuance of the resolution) we are reduced to an impossible construction, (such as the square root of a negative quantity), the case proposed is to be judged impossible. Which is yet here discovered to be otherwise.
>
> As for instance, the equation $aaa - 7a = 6$ should have for its root
>
> $$\sqrt[3]{3 + \sqrt{-\frac{100}{27}}} + \sqrt[3]{3 - \sqrt{-\frac{100}{27}}},$$
>
> which should therefore be judged an impossible case. Yet hath it a real root, $a = 3$, beside which it hath also two negatives, $a = -1$ and $a = -2$. And if we change but the sign of the absolute number (which is the only even place not vacant), the equation
>
> $$aaa - 7a = -6$$
>
> will have two affirmative roots, $a = 1$, $a = 2$ (and one negative root, $a = -3$).

Wallis's example, the equation $aaa - 7a = -6$, should look familiar. It is the one we used in Exercise 3.12 to illustrate the appearance of complex numbers in Cardano's formula.

Half a century later, in the 1739 paper "On the reduction of radicals to simpler terms, or the extraction of roots of any binomial $a + \sqrt{+b}$ or $a + \sqrt{-b}$," de Moivre sets himself the problem [17], [60, p. 447] "to extract the cube root of the impossible binomial $a + \sqrt{-b}$." He reduces this problem to solving a reduced cubic equation, but in a different way from how we proceeded in Exercises 4.13 and 4.14. Rather than quoting from his entirely readable text [60, p. 447], let's reproduce his approach as an exercise.

Exercise 5.34. Let a and b be real numbers with $b \neq 0$ and suppose $x + yi$ is a cube root of $a + bi$.

(i) Set $(x + yi)^3 = a + bi$, as in Exercise 4.13, and obtain two equations in x and y by setting the real and imaginary parts equal to each other.

(ii) Square both equations, take their sum, and obtain the equation

$$x^6 + 3x^4y^2 + 3x^2y^4 + y^6 = a^2 + b^2.$$

The left side is $(x^2 + y^2)^3$.

(iii) Let m be the real cube root of $a^2 + b^2$ and take cube roots of both sides to obtain $x^2 + y^2 = m$.

(iv) Substitute $m - x^2$ for y^2 in one of the equations of the first part to get a reduced cubic equation in x:

$$4x^3 - 3mx - a = 0.$$

(v) Calculate the discriminant of $(4x^3 - 3mx - a)/4$ and use the fact that $b \neq 0$ to show that it is positive.

(vi) Conclude that cube roots of complex numbers can be calculated by solving reduced cubic equations in the irreducible case.

Following de Moivre, we have converted the calculation of the cube root of a complex number to the solving of a reduced cubic equation of irreducible type. At this point, de Moivre observes that "this will best be done by means of a table of sines," illustrating the process by determining the cube roots of $81 + \sqrt{-2700}$. The cubic equation to be solved is

$$4x^3 - 63x - 81 = 0.$$

De Moivre suggests that this be "compared with the equation for the cosines, namely $4x^3 - 3r^2x = r^2c$." To understand the sense in which this is an equation for cosines, we can rewrite it as

$$4\left(\frac{x}{r}\right)^3 - 3\frac{x}{r} = \frac{c}{r}.$$

Taking x/r to be $\cos\theta$ and c/r to be $\cos 3\theta$ for a suitable angle θ, the equation becomes the trigonometric identity

$$4\cos^3\theta - 3\cos\theta = \cos 3\theta$$

of Theorem 5.12.

Following de Moivre again, we use the comparison of $4x^3 - 63x = 81$ to $4x^3 - 3r^2x = r^2c$ to obtain the values $r = \sqrt{21}$ and $c = 27/7$. To find x, we use the triple cosine identity and the identifications $x/r = \cos\theta$ and $c/r = \cos 3\theta$. Applying the inverse cosine to c/r, dividing by 3, computing the cosine, and multiplying by r yields x. De Moivre does the equivalent, thereby obtaining the equation's three solutions. From these, he determines three cube roots of $81 + \sqrt{-2700}$. (See [60, pp. 447–449] for the complete discussion.)

In our approach to solving reduced cubic equations, we have relied on Cardano's formula to produce a number whose cube root we need to calculate, and in the irreducible case applied de Moivre's formula to make the cube root calculation. In his 1739 paper, de Moivre has reversed the flow. He starts from a complex number whose cube root we want to calculate, produces a reduced cubic equation we wish to solve that falls under the irreducible case, then takes Viète's approach to solving the equation. De Moivre concludes his discussion with the observation that the use of trigonometry is essential [60, p. 449]:

> There have been several authors, and among them the eminent Wallis, who have thought that those cubic equations which are referred to the circle, may be solved by the extraction of the cube root of an imaginary quantity, and of $81 + \sqrt{-2700}$, without regard to the table of sines, but that is a mere fiction and a begging of the question. For on attempting it, the result always recurs back again to the same question as that first proposed. And the thing cannot be done directly, without the help of the table of sines.

He recognizes that computing cube roots of complex numbers and solving cubic equations in the irreducible case are essentially the same problem. We can loop back and forth between the two, but to break the cycle, we must introduce trigonometry.

Let us close our historical account of the cubic by examining Euler's treatment in *Elements of Algebra* [26]. Euler derives Cardano's formula in Chapter XII of Section 4, "Of the Rule of Cardan, or of Scipio Ferreo." His approach is simple and elegant, relying on the introduction of an associated quadratic polynomial. Compare the following excerpt [26, pp. 263–264] with Theorems 5.2 and 5.3 (ignoring the regrettable clash of notation).

> Whenever we meet with an equation of the form $x^3 = 3x\sqrt[3]{pq} + p + q$, we know that one of the roots is $\sqrt[3]{p} + \sqrt[3]{q}$. Now, we can determine p and q, in such a manner, that both $3\sqrt[3]{pq}$ and $p + q$ may be quantities equal to determinate numbers; so that we can always resolve an equation of the third degree, of the kind which we speak of [i.e., reduced].
>
> Let, in general, the equation $x^3 = fx + g$ be proposed. Here, it will be necessary to compare f with $3\sqrt[3]{pq}$ and g with $p + q$; that is, we

must determine p and q in such a manner, that $3\sqrt[3]{pq}$ may become equal to f, and $p + q = g$; for we then know that one of the roots of our equation will be $x = 3\sqrt[3]{p} + 3\sqrt[3]{q}$.

Following his derivation, Euler presents a sequence of examples of cubic equations with negative discriminant, starting with two for which the cube root calculations are trivial. He introduces a third example [26, p 265], $x^3 = 6x + 40$, with the immediate comment that "$x = 4$ is one of the roots." Applying Cardano's formula, he expresses the same surprise we did after Exercise 3.9 on comparing the complicated answer it produces with the known simpler answer:

> Consequently one of the roots will be ...
>
> $$x = \sqrt[3]{20 + 14\sqrt{2}} + \sqrt[3]{20 - 14\sqrt{2}};$$
>
> which quantity is really = 4, although, upon inspection, we should not suppose it. In fact, the cube of $2 + \sqrt{2}$ being $20 + 14\sqrt{2}$, we have, reciprocally, the cube root of $20 + 14\sqrt{2}$ equal to $2 + \sqrt{2}$; in the same manner, $\sqrt[3]{20 - 14\sqrt{2}} = 2 - \sqrt{2}$; wherefore our root $x = 2 + \sqrt{2} + 2 - \sqrt{2} = 4$.

Next Euler solves a non-reduced cubic, discussing the need (in Hewlett's translation) to "destroy" the second term and illustrating how to do so. The resulting reduced cubic is treated like the example above, after which Euler observes [26, p 268]:

> It was, however, by chance, as we have remarked, that we were able, in the preceding example, to extract the cube root of the binomials that we obtained, which is the case only when the equation has a rational root; ... But when there is no rational root, it is, on the other hand, impossible to express the root which we obtain in any other way, than according to the rule of Cardan; so that it is then impossible to apply reductions. For example, in the equation $x^3 = 6x + 4$, we have ... $x = \sqrt[3]{2 + 2\sqrt{-1}} + \sqrt[3]{2 - 2\sqrt{-1}}$, which cannot be otherwise expressed.

Thus, on arriving at the irreducible case, Euler declares that we can go no further than to leave the solution in such a complicated form.

In contrast to Cardano, Euler was comfortable with the use of complex numbers. We already quoted a passage from earlier in his book [26, p. 43] in which he wrote that "nothing prevents us from making use of these imaginary numbers, and employing them in calculation." Perhaps he chose not to pursue the example further because he preferred to avoid trigonometry as a tool. In a footnote [26, p 268] to Euler's passage, Bernoulli explains that the cubic is an example of

> the well-known *irreducible case*; a case which is so much the more remarkable, as the three roots are then always real. We cannot here make use of Cardan's formula, except by applying the methods of approximation. ... In the present work of Euler, we are not to look for all that might have been said on the direct and approximate resolutions of equations. He had too many curious and important objects, to dwell long upon this.

We have dwelled long upon this indeed. It is time to move beyond cubics.

6

Quartic Polynomials

We learned in Section 3.5 that Lodovicio Ferrari discovered a way to solve quartic equations not long after Cardano's work on cubic equations, and that Cardano presented Ferrari's method in *Ars Magna* [12]. We will take a brief look at Ferrari's approach, then turn to the approach of René Descartes. Both reduce the solving of a reduced quartic equation to the determination of the roots of an auxiliary cubic polynomial. A closer look at Descartes' solution will allow us to obtain a formula due to Euler that expresses the roots of the reduced quartic in terms of the roots of the auxiliary cubic. In turn, Euler's solution leads to a formula for the discriminant of a quartic polynomial. We will then have our final look at the effectiveness of coefficients as a tool to glean information about a polynomial's roots.

6.1 Reduced Quartics

From our study of graphs of polynomials in Section 1.6, we have an idea of what the graph of a quartic polynomial looks like and the number of roots it may have. Assuming the quartic is monic, its graph will fall from infinity, turn one or three times, then rise to infinity. The number of real roots will depend on the number of turning points and their location above, on, or below the x-axis. For instance, if there is one turning point, then we are in a situation similar to that of quadratic polynomials, with two, one, or no real roots depending on whether the turning point is below, on, or above the x-axis. If there are three turning points, there are many possibilities. We will not discuss them all, satisfying ourselves with those that can be illustrated through one family of examples. The interested reader can use them as a starting point and complete a survey of the options.

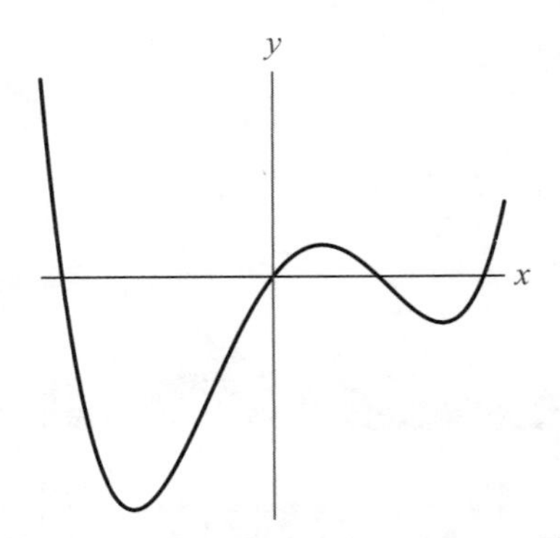

Figure 6.1. Graph of $y = x^4 - x^3 - 4x^2 + 4x$

Consider the infinite family of quartic polynomials $x^4 - x^3 - 4x^2 + 4x + e$, where e takes on all real number values. Figure 6.1 depicts the graph of $y = x^4 - x^3 - 4x^2 + 4x$, corresponding to $e = 0$. We see from the graph that this polynomial has four distinct real roots. Or, we see it directly from

$$x^4 - x^3 - 4x^2 + 4x = x(x + 2)(x - 1)(x - 2).$$

Imagine allowing e to increase from negative values of e with large absolute value to 0 and on through positive values of e. As e grows, the graph moves upward in the plane. For a negative value of e that is large in absolute value, the graph will meet the x-axis in two points and the quartic has two real roots. As e increases, it assumes the value for which the graph's one local maximum is on the x-axis, the case of three real roots. As e increases further, the local maximum rises above the x-axis, which then intersects the graph in four points, as illustrated in Figure 6.1. This corresponds to the existence of four real roots. Then e assumes the value at which the higher of the two local minima makes contact with the x-axis, giving us once again a quartic with three real roots. As e continues to increase, the higher local minimum rises above the x-axis and we have just two real roots. This remains the case until e assumes the value at which the lower local minimum makes contact with the x-axis. In this case, the quartic has one real root. For larger e the graph lies entirely above the x-axis and the quartic has no real roots.

Before analyzing quartic polynomials in general, we need some elementary facts that are similar to those we obtained to begin our treatment of cubics.

Exercise 6.1. Given A and B, we know that

$$(A + B)^2 = A^2 + 2AB + B^2$$

and

$$(A+B)^3 = A^3 + 3A^2B + 3AB^2 + B^3.$$

Show that

$$(A+B)^4 = A^4 + 4A^3B + 6A^2B^2 + 4AB^3 + B^4.$$

A quartic polynomial of the form

$$z^4 + qz^2 + rz + s$$

is called a *reduced* quartic.

Exercise 6.2. Let

$$f(x) = x^4 + bx^3 + cx^2 + dx + e,$$

for real numbers b, c, d, and e. Given another real number a, let's see what happens under the change of variable $x = z + a$, or $z = x - a$.

(i) Substitute $z + a$ for x and obtain a polynomial $g(z)$ in the new variable z. Write it as

$$z^4 + Bz^3 + Cz^2 + Dz + E$$

and obtain formulas for the coefficients B, C, D, and E in terms of a and the coefficients b, c, d, and e.

(ii) There is a choice of a for which $B = 0$, giving a new polynomial $g(z)$ that is reduced. Verify that $g(z)$ is the reduced polynomial described in Theorem 6.1.

(iii) What is the relation between a root of $f(x)$ and a root of $g(z)$?

(iv) Conclude that solving $g(z) = 0$ gives solutions of $f(x) = 0$.

Theorem 6.1. *Let b, c, d, and e be real numbers. Given a quartic polynomial $x^4 + bx^3 + cx^2 + dx + e$, the change of variable $x = z - b/4$ yields the reduced quartic polynomial*

$$z^4 + \left(-\frac{3b^2}{8} + c\right) z^2 + \left(\frac{b^3}{8} - \frac{bc}{2} + d\right) z + \left(-\frac{3b^4}{256} + \frac{b^2c}{16} - \frac{bd}{4} + e\right).$$

Exercise 6.3. What is the reduced quartic equation associated to the quartic equation

$$x^4 + 4x^3 - 2x + 4 = 0?$$

How are the solutions of the reduced quartic equation related to the solutions of the original quartic equation?

6.2 Ferrari's Method

A method for solving quartic equations was first discovered Lodovico Ferrari. In recognition of the historical importance of Ferrari's approach, we devote this section to it. However, we will not use his approach in the remainder of the chapter. Thus, this section can be omitted.

Suppose that we wish to solve the reduced quartic equation

$$z^4 + qz^2 + rz + s = 0$$

with real coefficients q, r, and s. Moving the lower-degree terms to the right to obtain

$$z^4 + qz^2 = -rz - s,$$

we can regard the left side as a quadratic polynomial in z^2 and complete the square by adding $qz^2 + q^2$ to both sides. This yields

$$(z^2 + q)^2 = qz^2 + q^2 - rz - s.$$

Ferrari's idea is to introduce an auxiliary term t, which is best thought of not as a variable but as a to-be-determined real number, and use it to change the left side from $(z^2 + q)^2$ to $(z^2 + q + t)^2$. The goal then becomes to find a value of t that makes the right side a square as well, allowing us to take square roots of both sides and reduce to a simpler problem. Let's do this.

To begin, we add $2(z^2 + q)t + t^2$ to both sides of $(z^2 + q)^2 = qz^2 + q^2 - rz - s$. The left side becomes a square and we obtain the equation

$$(z^2 + q + t)^2 = qz^2 + q^2 - rz - s + 2(z^2 + q)t + t^2.$$

Keep in mind that q, r, and s are specific real numbers, as is t, although we don't yet know what t is. The only variable is z. It is natural, then, to collect terms on the right side of the equation so that it is exhibited as a quadratic polynomial in z with coefficients involving the constants q, r, s, and t:

$$(z^2 + q + t)^2 = (q + 2t)z^2 - rz + (q^2 - s + 2qt + t^2).$$

Our goal is to find a value of t for which the right side becomes the square of a linear polynomial in z.

We know that a monic quadratic polynomial $z^2 + bz + c$ is the square of a linear polynomial precisely when it has a real root of multiplicity 2, or equivalently, when its discriminant $b^2 - 4c$ is 0. A polynomial of the form

$az^2 + bz + c$, with $a \neq 0$, will factor in the same way as $z^2 + (b/a)z + c/a$. Thus, it will be the square of a linear polynomial precisely when $(b/a)^2 - 4(c/a) = 0$, or $b^2 - 4ac = 0$.

We can apply this to

$$(q + 2t)z^2 - rz + (q^2 - s + 2qt + t^2)$$

viewed as a quadratic polynomial in z. It will be the square of a linear polynomial in z if t is chosen so that

$$(-r)^2 - 4(q + 2t)(q^2 - s + 2qt + t^2) = 0.$$

Rewriting the expression on the left side as a cubic polynomial in the unknown constant t and multiplying by -1, we find that t must satisfy

$$8t^3 + (20q)t^2 + (16q^2 - 8s)t + (4q^3 - 4qs - r^2) = 0.$$

Choosing such a t ensures that the quadratic polynomial in z is the square of a degree-one polynomial, so that both sides of the earlier equation

$$(z^2 + q + t)^2 = (q + 2t)z^2 - rz + (q^2 - s + 2qt + t^2)$$

are squares.

Using the techniques we have learned for solving cubic equations, we can find t, then factor the quadratic in z as the square of a degree-one polynomial $dz + e$. In this way, we obtain the quartic equation

$$(z^2 + q + t)^2 = (dz + e)^2.$$

Taking square roots on both sides, we obtain quadratic equations

$$z^2 + q + t = dz + e$$

and

$$z^2 + q + t = -dz - e.$$

Solving for z yields four solutions to the original quartic equation. We (and Ferrari) have reduced the problem of solving the original quartic equation $z^4 + qz^2 + rz + s = 0$ to that of solving the cubic equation in t.

Let's try Ferrari's approach in an example, one we will pursue only to the point of obtaining the reduced cubic equation that needs solving.

Exercise 6.4. Solve the equation

$$z^4 + 6z^2 - 60z + 36 = 0.$$

(i) Use Ferrari's method to show this can be done by finding t satisfying

$$t^3 + 15t^2 + 36t - 450 = 0.$$

(ii) Change variables to transform the cubic equation to a reduced cubic equation,

$$u^3 - 39u - 380 = 0.$$

(iii) Find a solution using Cardano's formula.

(iv) Explain how to obtain a solution to the original quartic equation.

Let's work out the next example to conclusion. It is missing both degree three and degree two terms, so that the cubic equation to solve is already reduced.

Exercise 6.5. Solve the quartic equation

$$z^4 - 12z + 3 = 0.$$

(i) Following Ferrari's procedure with $q = 0$, $r = -12$, and $s = 3$, introduce a new variable t and show that the equation takes the form

$$(z^2 + t)^2 = 2tz^2 + 12z + (t^2 - 3).$$

(ii) Show that the right side of this equation will be the square of a degree-one polynomial in z if t is a solution of the cubic equation

$$8t^3 - 24t - 144 = 0,$$

or

$$t^3 - 3t - 18 = 0.$$

(iii) Use Cardano's formula to show that one solution is

$$t = \sqrt[3]{9 + 4\sqrt{5}} + \sqrt[3]{9 - 4\sqrt{5}}.$$

(iv) Use the method of Exercise 3.10 to guess that a cube root of $9 + 4\sqrt{5}$ has the form $a + b\sqrt{5}$, obtain two equations in a and b, and guess values of a and b, leading to the discovery that the two cube roots we want are $3/2 + \sqrt{5}/2$ and $3/2 - \sqrt{5}/2$. Alternatively, guess that there is some small positive integer solution to the cubic equation, without reference to Cardano's formula, and find it.

(v) Substitute t into the equation in z and t at the start of the exercise to obtain

$$(z^2 + 3)^2 = 6z^2 + 12z + 6.$$

(vi) As Ferrari's method ensures, the right side of this equation is the square of a degree-one polynomial in z:

$$6z^2 + 12z + 6 = 6(z + 1)^2.$$

(vii) Using this, write the equation as

$$(z^2 + 3)^2 = 6(z + 1)^2.$$

(viii) Take square roots on both sides to get

$$z^2 - \sqrt{6}z + (3 - \sqrt{6}) = 0$$

and

$$z^2 + \sqrt{6}z + (3 + \sqrt{6}) = 0.$$

(ix) Solve them for z, thereby obtaining four solutions to

$$z^4 - 12z + 3 = 0.$$

6.3 Descartes' Method

We turn in this section to an approach to solving reduced quartic equations introduced by René Descartes. The idea is to factor a reduced quartic polynomial $z^4 + qz^2 + rz + s$ as a product of quadratic polynomials with real coefficients. If we can do this, then we can find the quartic's roots by applying the quadratic formula to its quadratic factors.

Let's first treat a special case, one for which we will find the desired factorization, although the case is sufficiently simple that we could find the roots directly.

Exercise 6.6. Let q and s be real numbers and consider the polynomial $z^4 + qz^2 + s$.

(i) Assume that $q^2 - 4s \geq 0$. Show that $z^4 + qz^2 + s$ factors as $(z^2 + l)(z^2 + m)$.

(ii) Assume that $q^2 - 4s < 0$ and observe that this implies $s > 0$. Verify that a factorization of the form $(z^2 + l)(z^2 + m)$ can't exist. Therefore, to show that $z^4 + qz^2 + s$ factors as a product of quadratic polynomials with real coefficients, we will instead seek a factorization of the more general form

$$z^4 + qz^2 + s = (z^2 + kz + l)(z^2 + k'z + m),$$

where the coefficients k, k', l, and m are real and k and k' are nonzero. Our search will occupy the remainder of the exercise.

(iii) Multiply the two quadratic polynomials, obtain expressions for the coefficients of each power of z, and show that for equality to hold, we must have $k' = -k$ and $l = m$. Thus, we can simplify our task and look for a factorization of the form

$$z^4 + qz^2 + s = (z^2 + kz + l)(z^2 - kz + l).$$

(iv) Show next that for equality to hold, k and l must satisfy $l^2 = s$ and $2l = k^2 + q$. We view these as equations in the unknowns k and l, with q and s regarded as known constants.

(v) Square both sides of $2l = k^2 + q$, substitute s for l^2, and show that k must satisfy

$$k^4 + 2qk^2 + (q^2 - 4s) = 0.$$

(vi) This is a quadratic equation in k^2. Set $j = k^2$ and write it as

$$j^2 + 2qj + (q^2 - 4s) = 0.$$

(vii) Check that the discriminant of $j^2 + 2qj + (q^2 - 4s)$ as a quadratic polynomial in j is $16s$, which is positive. Deduce that the quadratic polynomial's roots are real and distinct. Using the assumption that $q^2 - 4s < 0$, deduce that there is a positive real root. (Recall Theorem 5.13.)

(viii) Let k and $-k$ represent the square roots of the positive root. Conclude that k and $-k$ give a value for l and allow us to factor $z^4 + qz^2 + s$ as a product of quadratic polynomials with real coefficients.

Exercise 6.6 shows that a quartic polynomial of the form $z^4 + qz^2 + s$ factors as a product of quadratic polynomials. We can use the factorization to find the roots of $z^4 + qz^2 + s$. Alternatively, we can find them directly.

Exercise 6.7. Use the quadratic formula to show that the roots of $z^4 + qz^2 + s$ are

$$\pm\sqrt{-\frac{q}{2} \pm \frac{\sqrt{q^2 - 4s}}{2}}.$$

The argument in Exercise 6.6 showing when $q^2 - 4s < 0$ that $z^4 + qz^2 + s$ can be factored as a product of quadratic polynomials with real coefficients can be extended to quartic polynomials with non-zero linear term.

Exercise 6.8. Let q, r, and s be real numbers, with $r \neq 0$, and consider the polynomial $z^4 + qz^2 + rz + s$.

(i) We wish to find real numbers k, k', l, and m for which $z^4 + qz^2 + rz + s$ factors as

$$(z^2 + kz + l)(z^2 + k'z + m).$$

Show that for this to hold, we must have $k' = -k$. Thus we can look for a factorization of the form

$$z^4 + qz^2 + rz + s = (z^2 + kz + l)(z^2 - kz + m).$$

Check also, since $r \neq 0$, that k must be non-zero.

(ii) Multiply the two quadratic polynomials and show that for equality to hold, k, l, and m must satisfy

$$l + m - k^2 = q; \;\; k(m - l) = r; \;\; lm = s.$$

(iii) Since k can't be zero, we are free to divide by it. Do so in the second equation and obtain from the first two equations the new equations

$$m + l = q + k^2; \;\; m - l = \frac{r}{k}.$$

Using these, obtain

$$2m = q + k^2 + \frac{r}{k}; \;\; 2l = q + k^2 - \frac{r}{k}.$$

Conclude that l and m are determined once k is known.

(iv) We can multiply both sides of the equation $lm = s$ by 4 to obtain $(2l)(2m) = 4s$. Substitute the expressions for $2l$ and $2m$ and obtain

$$k^6 + 2qk^4 + (q^2 - 4s)k^2 - r^2 = 0.$$

(v) This is a cubic equation in k^2. Set $j = k^2$ and write it as

$$j^3 + 2qj^2 + (q^2 - 4s)j - r^2 = 0.$$

(vi) Deduce from Theorem 5.17 that the cubic equation in j has a positive real solution.

(vii) Let k and $-k$ represent the square roots of this solution. Conclude that k and $-k$ give values for l and m and allow us to factor $z^4 + qz^2 + rz + s$ as a product of quadratic polynomials with real coefficients. The roots of the quadratic polynomials yield the solutions of the original quartic equation.

(viii) If $s = 0$, then solving the original quartic equation reduces to solving $z^3 + qz + r = 0$. The degree-six polynomial in k becomes $k^6 + 2qk^4 + q^2k^2 - r^2$, which factors as $(k^3 + qk + r)(k^3 + qk - r)$. Hence, setting it equal to 0 and solving for k is essentially the same as solving the original equation.

We have obtained a procedure for finding roots of quartic polynomials. First we pass from a quartic polynomial

$$x^4 + bx^3 + cx^2 + dx + e$$

to its associated reduced quartic polynomial

$$z^4 + qz^2 + rz + s.$$

If $r \neq 0$, we write the cubic polynomial

$$j^3 + 2qj^2 + (q^2 - 4s)j - r^2,$$

which is called the *resolvent cubic* of the quartic $z^4 + qz^2 + rz + s$. If $r = 0$ and $q^2 - 4s < 0$, we can work with the same cubic, or its quadratic factor $j^2 + 2qj + (q^2 - 4s)$. We find a positive real root of the resolvent cubic, take its square roots, use them to factor the reduced quartic as a product of quadratic polynomials with real coefficients, find their roots, and thereby obtain the roots of the original quartic polynomial. If $r = 0$ and $q^2 - 4s \geq 0$, we obtain a factorization of $z^4 + qz^2 + rz + s$ directly as a product of quadratics, using the quadratic formula.

We restate some of this discussion in the next theorem.

Theorem 6.2. *Let $z^4 + qz^2 + rz + s$ be a reduced quartic polynomial with real coefficients. Assume that $r \neq 0$ or that $r = 0$ and $q^2 - 4s < 0$. Let*

$$j^3 + 2qj^2 + (q^2 - 4s)j - r^2$$

be the given quartic's associated resolvent cubic polynomial and let k and $-k$ be square roots of a positive real root of the resolvent cubic. Then $z^4 + qz^2 + rz + s$ factors as

$$\left(z^2 + kz + \frac{1}{2}\left(q + k^2 - \frac{r}{k}\right)\right)\left(z^2 - kz + \frac{1}{2}\left(q + k^2 + \frac{r}{k}\right)\right).$$

The roots of $z^4 + qz^2 + rz + s$ are the roots of the two quadratic polynomials in the factorization.

The procedure for finding roots of quartic polynomials can be long and complicated, but in principle it works. The hardest part is likely to be the calculation of roots of the resolvent cubic, a problem we already understand.

Let's try the procedure on some examples. They are designed so that the polynomials arising as resolvent cubics have positive real roots that are easy to find. Thus, in solving the quartic equations, we can focus on the new features of the procedure and not get distracted by the now-familiar difficulties of the cubic.

Exercise 6.9. Find the four solutions of

$$z^4 - 3z^2 + 6z - 2 = 0.$$

To do so, write the resolvent cubic, find a real root by guessing (squares of integers are good guesses), take its two square roots, factor $z^4 - 3z^2 + 6z - 2$ as a product of two quadratic polynomials, and find their roots.

Exercise 6.10. Find the four solutions of

$$z^4 - 2z^2 - 8z - 3 = 0$$

by following the procedure outlined in Exercise 6.9.

Our third example is the quartic equation of Exercise 6.5. We will solve it using Descartes' method.

Exercise 6.11. Find the four solutions of

$$z^4 - 12z + 3 = 0.$$

In this case, since there is no square term, the resolvent cubic will be a reduced cubic, so that Cardano's formula can be applied directly. Find a real root by using Cardano's formula or by guessing. (Guessing may not lead to a root that is the square of an integer, but it may lead to a positive integer.) Then proceed as in the preceding exercises. Compare the solutions to the result in Exercise 6.5 to make sure they are the same.

We have found that a reduced quartic polynomial

$$z^4 + qz^2 + rz + s$$

with real coefficients factors as a product of quadratic polynomials with real coefficients. From this, we can obtain information on the nature of the roots of any quartic polynomial.

Theorem 6.3. *Any quartic polynomial $f(x)$ with real coefficients factors as a product of quadratic polynomials with real coefficients. Exactly one of the following possibilities occurs for the roots of $f(x)$:*

(i) $f(x)$ has four distinct real roots.

(ii) $f(x)$ has two distinct real roots and a pair of non-real complex conjugate roots.

(iii) $f(x)$ has no real roots, but two pairs of distinct non-real complex conjugate roots.

(iv) $f(x)$ has repeated roots.

Exercise 6.12. Deduce Theorem 6.3 for reduced quartic polynomials from Exercise 6.6 and Theorem 6.2. Then explain why it holds for quartic polynomials in general.

6.4 Euler's Formula

We can refine Descartes' method of solving a reduced quartic equation (Exercise 6.8 and Theorem 6.2) to obtain a formula for its solutions in terms of the solutions of the associated resolvent cubic equation. The formula is due to Euler, who obtained it directly rather than as a consequence of Descartes' result. We will be content to derive it from Descartes' factorization. Euler's approach will be discussed at the end of Section 6.8.

We consider once again the reduced quartic equation

$$z^4 + qz^2 + rz + s = 0.$$

In light of the explicit description of its solutions in Exercise 6.7 when $r = 0$, let's assume that $r \neq 0$. Suppose the three solutions of the resolvent cubic equation

$$j^3 + 2qj^2 + (q^2 - 4s)j - r^2 = 0.$$

are j_1, j_2, and j_3. Our goal is to describe the roots of $z^4 + qz^2 + rz + s$ in terms of j_1, j_2, and j_3.

We have already shown that at least one of the resolvent cubic's roots is positive and real. Let's assume it is j_1, and let k_1 and $-k_1$ be its square roots. Because j_1 is positive, k_1 and $-k_1$ are real. The other two roots of the resolvent cubic, j_2 and j_3, may be real or may be complex conjugates.

Whether or not j_2 and j_3 are real, their square roots are opposites of each other. Thus we can write k_2 and $-k_2$ for the square roots of j_2 and k_3 and $-k_3$ for the square roots of j_3.

Do not be fooled by the notation. Seeing the pair k_i and $-k_i$, you should not assume that k_i is positive or that $-k_i$ is negative. In fact, they may not even be real. The notation should convey only that whatever they are, each is the opposite of the other.

Exercise 6.13. Use the notation and assumptions of the preceding discussion.

(i) Explain why there are eight possible choices for the triple of numbers (k_1, k_2, k_3), depending on choice of signs.

(ii) Show that

$$j_1 + j_2 + j_3 = -2q, \quad j_1 j_2 j_3 = r^2.$$

(Hint: Use Theorem 5.14 on how the roots and coefficients of a cubic polynomial are related.)

(iii) Deduce that the product $k_1 k_2 k_3$ equals either r or $-r$, depending on which square roots of the j_i's are chosen as the k_i's. More precisely, deduce that four of the eight choices of (k_1, k_2, k_3) result in $k_1 k_2 k_3 = r$ and that the other four choices result in $k_1 k_2 k_3 = -r$.

Exercise 6.14. Continue within the setting of Exercise 6.13. For definiteness, let's fix the choice of triple of square roots (k_1, k_2, k_3) to be one of the four satisfying $k_1 k_2 k_3 = -r$.

(i) Given our choice of k_1 as one of the square roots of the positive real number j_1, we obtain from Theorem 6.2 the factorization of $z^4 + qz^2 +$

$rz + s$ as

$$\left(z^2 + k_1 z + \frac{1}{2}\left(q + k_1^2 - \frac{r}{k_1}\right)\right)\left(z^2 - k_1 z + \frac{1}{2}\left(q + k_1^2 + \frac{r}{k_1}\right)\right).$$

Use the quadratic formula to obtain expressions for the roots of

$$z^2 + k_1 z + \frac{1}{2}\left(q + k_1^2 - \frac{r}{k_1}\right).$$

Show that

$$z = -\frac{k_1}{2} \pm \frac{1}{2}\sqrt{k_1^2 - 2\left(q + k_1^2 - \frac{r}{k_1}\right)}.$$

(ii) Use the formulas from Exercise 6.13 for $-2q$ and r in terms of the k_i's as well as the choice of the triple (k_1, k_2, k_3) to satisfy $k_1 k_2 k_3 = -r$ to write this equality as

$$z = -\frac{k_1}{2} \pm \frac{1}{2}\sqrt{(k_2 - k_3)^2}.$$

(iii) Conclude that the roots of the quadratic factor are

$$z = \frac{1}{2}(-k_1 + k_2 - k_3) \quad \text{and} \quad z = \frac{1}{2}(-k_1 - k_2 + k_3).$$

(iv) Follow the same procedure to find the roots of

$$z^2 - k_1 z + \frac{1}{2}\left(q + k_1^2 + \frac{r}{k_1}\right)$$

and show that you get

$$z = \frac{1}{2}(k_1 + k_2 + k_3) \quad \text{and} \quad z = \frac{1}{2}(k_1 - k_2 - k_3).$$

(v) Conclude that once the three roots j_1, j_2, j_3 of the resolvent cubic polynomial are found, the roots of the quartic polynomial can be expressed in terms of square roots of j_1, j_2, and j_3, as described in Theorem 6.4.

Theorem 6.4. *Let q, r, and s be real numbers with $r \neq 0$. Let j_1, j_2, and j_3 be the roots of the resolvent cubic polynomial*

$$j^3 + 2qj^2 + (q^2 - 4s)j - r^2$$

associated with the quartic polynomial

$$z^4 + qz^2 + rz + s.$$

Then the roots of $z^4 + qz^2 + rz + s$ *are the four sums*

$$\frac{1}{2}\left(\sqrt{j_1} + \sqrt{j_2} + \sqrt{j_3}\right)$$

arising from the choices of square roots $\sqrt{j_1}$, $\sqrt{j_2}$, $\sqrt{j_3}$ *that satisfy*

$$\sqrt{j_1} \cdot \sqrt{j_2} \cdot \sqrt{j_3} = -r.$$

Let's solve two quartic equations using Theorem 6.4. The equations have been chosen so that the roots of the resolvent cubic polynomials are easily found by trial and error.

Exercise 6.15. Find all four solutions of

$$z^4 - 3z^2 + \sqrt{6}z - \frac{1}{2} = 0$$

and all four solutions of

$$z^4 - 3z^2 - \sqrt{6}z - \frac{1}{2} = 0.$$

In the last exercise, the two quartic equations have the same resolvent cubic equation. This is a special case of a general phenomenon.

Exercise 6.16. Explain why the quartic equations

$$z^4 + qz^2 + rz + s = 0$$

and

$$z^4 + qz^2 - rz + s = 0$$

have the same resolvent cubic equation. Describe how the solutions of the first are related to the solutions of the second.

6.5 The Discriminant

The discriminant of a quadratic or cubic polynomial is defined as the square of the product of differences of the roots. The discriminant of a quartic polynomial

$$x^4 + bx^3 + cx^2 + dx + e$$

is defined in the same way. Suppose its roots are r_1, r_2, r_3, and r_4. Then its *discriminant* Δ is the product

$$(r_1 - r_2)^2(r_1 - r_3)^2(r_1 - r_4)^2(r_2 - r_3)^2(r_2 - r_4)^2(r_3 - r_4)^2.$$

Because the root differences are squared, we get the same result regardless of how the roots are ordered.

For quadratic and cubic polynomials, the discriminant is important for two reasons. First, its sign gives us information on the nature of the roots: how many are real numbers and how many are non-real complex numbers. Second, there is a formula expressing the discriminant in terms of the coefficients of the polynomial, allowing us to calculate the discriminant from the coefficients and thereby obtain information about the roots from the coefficients alone.

We will show for a quartic polynomial that the discriminant gives information on the nature of the roots, although not as complete as in the quadratic and cubic cases, and that it can be calculated from the coefficients.

Exercise 6.17. In this exercise, we will find out what the discriminant of a quartic polynomial tells us about the nature of its roots. Let $f(x)$ be the polynomial, with discriminant Δ, and let r_1, r_2, r_3, and r_4 be its roots.

(i) Check that $\Delta = 0$ precisely when at least two of the roots coincide.

(ii) Assume in the remainder of the exercise that the four roots are distinct, so that $\Delta \neq 0$. From Theorem 6.3, there are three possibilities for the roots: all are real, two are real and two are non-real complex conjugates, or there are two pairs of non-real complex conjugates.

(iii) Show that if all four roots are real, then $\Delta > 0$.

(iv) Suppose two roots are real and two are complex conjugates. Show that $\Delta < 0$. (Hint: The product of a non-zero complex number and its conjugate is a positive real number. Name the four roots r, s, $a + bi$, and $a - bi$, where r, s, a, and b are real numbers. There are six root differences. Four of the six differences occur as pairs of conjugate complex numbers, so that their products are real. One of the remaining root differences is real and the other is pure imaginary. From this it follows that Δ is negative.)

(v) Suppose that the four roots occur in two pairs of non-real complex conjugates. Show that $\Delta > 0$. (Hint: Pair four of the differences so that they are conjugates and their products are real. The other two root differences should be pure imaginary.)

(vi) Show that the converses of the statements in (iii)–(v) hold; that is, prove Theorem 6.5. (Hint: As in Exercises 4.7 and 5.7, these follow from the statements themselves by an elementary logical argument.)

Theorem 6.5. *Let $f(x)$ be a polynomial of degree 4 with discriminant Δ.*

(i) If Δ is positive, then $f(x)$ has four distinct roots. Either they are all real or they are all non-real, in which case they form two complex conjugate pairs.

(ii) If $\Delta = 0$, then $f(x)$ has a root occurring with multiplicity at least 2.

(iii) If Δ is negative, then $f(x)$ has four distinct roots. Two are real and two form a complex conjugate pair.

When the discriminant is positive, Theorem 6.5 does not completely settle the nature of the roots. Theorem 6.9 will describe how to make further use of the coefficients to determine whether the roots are all real or all non-real.

We wish to calculate Δ in terms of the coefficients of the polynomial. We begin with the special case of a reduced quartic with no linear term.

Exercise 6.18. Let q and s be real numbers. Use the description of the roots of

$$z^4 + qz^2 + s$$

in Exercise 6.7 to calculate the discriminant, obtaining Theorem 6.6.

Theorem 6.6. *Let q and s be real numbers. The discriminant of z^4+qz^2+s is given by*

$$\Delta = 16s(q^2 - 4s)^2 = -128q^2s^2 + 16q^4s + 256s^3.$$

The key to computing the discriminant of a general reduced quartic polynomial is to relate it to the discriminant of the quartic's resolvent cubic.

Exercise 6.19. Let q, r, and s be real numbers, with $r \neq 0$.

(i) Using the notation we employed in Exercise 6.14, write the roots r_1, r_2, r_3, and r_4 of $z^4 + qz^2 + rz + s$ as halves of sums of $\pm k_i$. Show that the six root differences, up to sign, are

$$k_1 + k_2, \;\; k_1 + k_3, \;\; k_2 + k_3,$$

$$k_1 - k_2, \;\; k_1 - k_3, \;\; k_2 - k_3.$$

(ii) Deduce that

$$\Delta = (k_1^2 - k_2^2)^2 (k_1^2 - k_3^2)^2 (k_2^2 - k_3^2)^2.$$

(iii) The k_i's are the square roots of the solutions j_1, j_2, j_3 of the resolvent cubic equation. Using the j_i's, write the equation for Δ as

$$\Delta = (j_1 - j_2)^2 (j_1 - j_3)^2 (j_2 - j_3)^2.$$

Since the j_i's are by definition the roots of the resolvent cubic polynomial, the product on the right side is in fact the discriminant of the resolvent cubic.

Since the discriminant of $z^4 + qz^2 + rz + s$ $(r \neq 0)$ coincides with the discriminant of its resolvent cubic, we can compute it using our results on discriminants of cubics. Let's do this in a simple special case first.

Exercise 6.20. Let r and s be real numbers with $r \neq 0$.

(i) Check that the resolvent cubic of $z^4 + rz + s$ is the reduced cubic

$$j^3 - 4sj - r^2.$$

(ii) Use the formula for the discriminant of a reduced cubic polynomial $y^3 + py + q$ in terms of p and q to show that its discriminant is

$$256s^3 - 27r^4.$$

(iii) Deduce that $256s^3 - 27r^4$ is the discriminant of $z^4 + rz + s$ also.

Exercise 6.21. Compute the discriminants of the quartic polynomials below and say what you can about their number of real roots.

(i) $z^4 + z + 1$.

(ii) $z^4 + z - 1$.

(iii) $z^4 + z + s$. (The answer depends on s.)

We now obtain a formula for the discriminant of a general reduced quartic polynomial.

Exercise 6.22. Let q, r, and s be real numbers, with $r \neq 0$.

(i) The resolvent cubic of $z^4 + qz^2 + rz + s$ has coefficients $2q$, $q^2 - 4s$, and $-r^2$. Use the formula in Theorem 5.6 for the discriminant of $x^3 + bx^2 + cx + d$, substitute for b, c, and d, and obtain a complicated expression in q, r, and s for the discriminant of the resolvent cubic.

(ii) Conclude that it is also the discriminant of the reduced quartic polynomial $z^4 + qz^2 + rz + s$ with which we began.

(iii) Expand and simplify the expression in q, r, and s. Combine with Theorem 6.6 to conclude that Theorem 6.7 holds.

Theorem 6.7. *The discriminant Δ of the reduced quartic polynomial $z^4 + qz^2 + rz + s$ is given by*

$$\Delta = 144qr^2s - 128q^2s^2 - 4q^3r^2 + 16q^4s - 27r^4 + 256s^3.$$

Exercise 6.23. Compute the discriminants of the cubic polynomials below. What can be said about their number of real roots?

(i) $z^4 - 3z^2 + 6z - 2$.

(ii) $z^4 - 2z^2 - 8z - 3$.

(iii) $z^4 - 3z^2 + \sqrt{6}z - \frac{1}{2}$.

The discriminant of an arbitrary quartic polynomial can be computed, although the process is laborious.

Exercise 6.24. Consider the quartic polynomial

$$x^4 + bx^3 + cx^2 + dx + e.$$

From Exercise 6.2 and Theorem 6.1, the change of variable $x = y - (b/4)$ transforms $x^4 + bx^3 + cx^2 + dx + e$ into a polynomial of the form $z^4 + qz^2 + rz + s$.

(i) Verify, as in Exercise 5.10, that although the roots of $x^4 + bx^3 + cx^2 + dx + e$ and the roots of $z^4 + qz^2 + rz + s$ are not the same, their differences are.

(ii) Deduce that the discriminant of $x^4 + bx^3 + cx^2 + dx + e$ is the same as the discriminant of $z^4 + qz^2 + rz + s$.

(iii) Conclude that a formula for the discriminant of $x^4+bx^3+cx^2+dx+e$ in terms of b, c, d, and e can be found by substituting into the formula for the discriminant of $z^4 + qz^2 + rz + s$ the expressions given by Theorem 6.1 for q, r, and s in terms of b, c, d, and e.

(iv) Carry out this calculation, simplify, and conclude that Theorem 6.8 holds.

Theorem 6.8. *The discriminant Δ of the quartic polynomial $x^4 + bx^3 + cx^2 + dx + e$ is given by*

$$\begin{aligned}\Delta = {} & 18bcd^3 + 18b^3cde - 80bc^2de - 6b^2d^2e + 144cd^2e \\ & + 144b^2ce^2 - 128c^2e^2 - 192bde^2 + b^2c^2d^2 - 4b^3d^3 \\ & - 4c^3d^2 - 4b^2c^3e + 16c^4e - 27d^4 - 27b^4e^2 + 256e^3.\end{aligned}$$

6.6 The Nature of the Roots

We saw in Theorem 6.5 that the discriminant of a quartic polynomial gives us partial information on the nature of its roots. In this section, we will do better by supplementing the discriminant with other expressions in the coefficients. We know for example that a quartic with positive discriminant has either four distinct real roots or four distinct non-real roots occurring in complex conjugate pairs. We will refine this, then turn to a closer analysis of quartic polynomials with zero discriminant.

Exercise 6.25. Consider the quartic polynomial

$$z^4 + qz^2 + s$$

and assume its discriminant Δ is positive. We wish to determine, in terms of q and s, whether the roots are all real or all non-real.

(i) Use the fact that $\Delta > 0$ and the formula for Δ (Theorem 6.6) to deduce that $s > 0$.

(ii) Deduce that the roots of $z^4 + qz^2 + s$ are non-zero, and that they will all be real precisely when the roots of the quadratic polynomial $t^2 + qt + s$ are real and positive.

(iii) Using Theorem 5.13 and the fact that $s > 0$, deduce that this occurs precisely when $q < 0$ and $q^2 - 4s > 0$.

Next we examine reduced quartic polynomials with non-zero linear term.

Exercise 6.26. Consider the quartic polynomial

$$z^4 + qz^2 + rz + s.$$

Assume that its discriminant Δ is positive and that $r \neq 0$.

(i) Deduce, since Δ is also the discriminant of the resolvent cubic polynomial

$$j^3 + 2qj^2 + (q^2 - 4s)j - r^2,$$

that its roots j_1, j_2, and j_3 are real and distinct.

(ii) From Theorem 5.14, deduce that

$$j_1 j_2 j_3 = r^2.$$

(iii) Conclude that either all the roots j_i are positive or one is positive and the other two are negative.

(iv) From Theorem 5.16, deduce that the j_i are all positive precisely when

$$q < 0 \text{ and } q^2 - 4s > 0.$$

(v) If the j_i are all positive, then their square roots $\pm k_i$ are all real numbers. Conclude from Theorem 6.4 that the four roots of the quartic $z^4 + qz^2 + rz + s$ are all real.

(vi) Suppose j_1 is positive and j_2 and j_3 are negative. Conclude that the two square roots $\pm k_1$ are real while $\pm k_2$ and $\pm k_3$ are pure imaginary. Show in this case that the four roots of $z^4 + qz^2 + rz + s$ are all non-real complex numbers.

(vii) Conclude that the roots of $z^4 + qz^2 + rz + s$ are all real precisely when $q < 0$ and $q^2 - 4s > 0$.

Combining the results of Exercises 6.25 and 6.26, we obtain:

Theorem 6.9. *The roots of a reduced quartic polynomial* $z^4 + qz^2 + rz + s$ *with positive discriminant are all real if* $q < 0$ *and* $q^2 - 4s > 0$, *and all non-real otherwise.*

This result is due to Lagrange, who stated it in his *Traité de la Résolution des Équations Numériques de Tous les Degrés* [39, p. 68], which first appeared in 1798.

Exercise 6.27. For each of the quartic polynomials, use the discriminant and the coefficients to decide how many roots are real.

(i) $z^4 + z + 1$.

(ii) $z^4 + z - 1$.

(iii) $z^4 + z + s$. (The answer depends on s.)

(iv) $z^4 - 3z^2 + 6z - 2$.

(v) $z^4 - 2z^2 - 8z - 3$.

(vi) $z^4 - 3z^2 + \sqrt{6}z - \frac{1}{2}$.

A version of Theorem 6.9 for arbitrary quartic polynomials is easily derived. A polynomial $x^4 + bx^3 + cx^2 + dx + e$ has the same discriminant as its associated reduced quartic polynomial $z^4 + qz^2 + rz + s$. Moreover, the two are related through a change of variable by the real number $-b/4$, as described in Theorem 6.1. Hence, if their shared discriminant is positive, then both have all real roots or all non-real roots. Which of the two occurs, according to Theorem 6.9, is determined by the signs of q and $q^2 - 4s$, and these can be expressed in terms of b, c, d, and e, as in Theorem 6.1. This yields a theorem describing whether the roots of $x^4 + bx^3 + cx^2 + dx + e$ are all real or all non-real in terms of b, c, d, and e.

Combining this with Theorem 6.5, we can decide from a quartic polynomial's coefficients when it has roots that are real and distinct, non-real and distinct, or a mix of real and non-real but distinct. The remaining possibility, which occurs when the quartic has zero discriminant, is that of a repeated root. We conclude our development of the quartic with an examination of this case, leaving many of the details for the reader to work out.

Once again, we look first at quartics whose linear term is zero, recalling from Theorem 6.6 that $z^4 + qz^2 + s$ has discriminant $16s(q^2 - 4s)^2$. Hence, for the discriminant to be 0, either $s = 0$ or $q^2 - 4s = 0$.

Theorem 6.10. *Let $z^4 + qz^2 + s$ be a quartic polynomial with zero discriminant.*

(i) Assume $s = 0$. Then $z^4 + qz^2$ factors as $z^2(z^2 + q)$.

- *(a) If $q = 0$, then 0 is a root of multiplicity 4.*
- *(b) If $q \neq 0$, then 0 is a root of multiplicity 2 and the square roots of $-q$ are roots of multiplicity 1, real if $q < 0$ and pure imaginary if $q > 0$.*

(ii) Assume $s \neq 0$, so that $q^2 - 4s = 0$ and $q \neq 0$. Then $z^4 + qz^2 + s$ factors as $(z^2 + q/2)^2$.

- *(a) If $q < 0$, then $z^4 + qz^2 + s$ has two real roots of multiplicity 2, each the other's opposite.*
- *(b) If $q > 0$, then $z^4 + qz^2 + s$ has two pure imaginary roots of multiplicity 2, each the other's opposite.*

Exercise 6.28. Prove Theorem 6.10

To continue, we need the quartic analogue of Theorem 5.14, which describes the coefficients of a cubic polynomial in terms of its roots.

Theorem 6.11. *Let $x^4+bx^3+cx^2+dx+e$ be a quartic polynomial with real coefficients b, c, d, and e, and suppose its roots are r_1, r_2, r_3, and r_4. Then*

$$\begin{aligned} b &= -r_1 - r_2 - r_3 - r_4, \\ c &= +r_1r_2 + r_1r_3 + r_1r_4 + r_2r_3 + r_2r_4 + r_3r_4, \\ d &= -r_1r_2r_3 - r_1r_2r_4 - r_1r_3r_4 - r_2r_3r_4, \\ e &= +r_1r_2r_3r_4. \end{aligned}$$

Exercise 6.29. Prove Theorem 6.11. (Hint: Use the factorization of $x^4 + bx^3 + cx^2 + dx + e$ as $(x - r_1)(x - r_2)(x - r_3)(x - r_4)$.)

Theorem 6.11 tells us in particular that the sum of the roots of a reduced quartic polynomial is 0. We can now complete our analysis of the roots of a such a polynomial when it has zero discriminant.

Theorem 6.12. *Let $z^4 + qz^2 + rz + s$ be a reduced quartic polynomial with zero discriminant.*

(i) *If $z^4 + qz^2 + rz + s$ has a root of multiplicity 4, or two roots of multiplicity 2, or 0 as a root of multiplicity 2, then $r = 0$ and the roots are as described in Theorem 6.10.*

(ii) *Assume $r \neq 0$. Then $z^4 + qz^2 + rz + s$ has either a non-zero real root of multiplicity 3 and a real root of multiplicity 1 or a non-zero real root of multiplicity 2 and two roots of multiplicity 1.*

  (a) *If $q < 0$ and $q^2 - 4s > 0$, the roots are all real. If $q^2 + 12s = 0$, there is a non-zero real root a of multiplicity 3 and $-3a$ is a root of multiplicity 1. If $q^2 + 12s \neq 0$, there is a non-zero real root a of multiplicity 2 and there are two real roots of the form $-a + c$ and $-a - c$ for some non-zero real number c with $c \neq \pm 2a$.*

  (b) *If it is not the case that $q < 0$ and $q^2 - 4s > 0$, then there is a non-zero real root a of multiplicity 2 and there are two complex conjugate roots of the form $-a + bi$ and $-a - bi$, for some non-zero real number b.*

Exercise 6.30. Prove Theorem 6.12.

(i) From Theorem 6.3, if a non-real complex number occurs as a quartic's root, so does its conjugate. Using this and Theorem 6.11, prove (i) by calculating r in terms of the roots. (Remember, the roots add to 0.)

(ii) Assume $r \neq 0$. Show that there is a non-zero real root a of multiplicity 2 or 3.

(iii) Show that if all the roots are real, they have the form $a, a, -a+c, -a-c$ for some non-zero real number c, but if they are not all real, they have the form $a, a, -a+bi, -a-bi$ for some non-zero real number b.

(iv) Use Theorem 6.11 to calculate q, r and s in terms of the roots. Verify that $q < 0$ and $q^2 - 4s > 0$ if the roots are all real, but not otherwise.

(v) Compute $q^2 + 12s$ in terms of the roots and show that it is the square of a simple expression. Check that it can't be 0 unless the roots are all real, in which case it is 0 precisely when $c = \pm 2a$. This is exactly the case of a multiplicity 3 root.

In the case of a quartic polynomial $z^4 + qz^2 + rz + s$ with zero discriminant and $r \neq 0$, Theorem 6.12 focuses attention on the quantity $q^2 + 12s$. Whether it is zero or not determines whether the polynomial has a triple root or not. A similar result holds for $z^4 + qz^2 + s$: for its discriminant to be zero, either $s = 0$ or $q^2 - 4s = 0$. Either way, if $q^2 + 12s = 0$, then $q = s = 0$, and conversely. Thus, $q^2 + 12s = 0$ precisely when $z^4 + qz^2 + s$ has a quadruple root. This yields the following theorem.

Theorem 6.13. *The polynomial $z^4 + qz^2 + rz + s$ has a root of multiplicity 3 or 4 precisely when $\Delta = 0$ and $q^2 + 12s = 0$.*

We can use Theorem 6.1 on change of variable to extend our results on reduced quartic polynomials with zero discriminant to arbitrary quartic polynomials with zero discriminant. As one example, we obtain:

Theorem 6.14. *The polynomial $x^4 + bx^3 + cx^2 + dx + e$ has a root of multiplicity 3 or 4 precisely when $\Delta = 0$ and $c^2 - 3bd + 12e = 0$.*

Exercise 6.31. Prove Theorem 6.14.

These theorems provide our final illustration of the theme that a polynomial's coefficients encode rich information on its roots.

6.7 Cubic and Quartic Reprise

We close our treatment of cubic and quartic polynomials with three exercises that provide an opportunity to review what we have learned. In each exercise, the reader is given a list of items to address and asked to write an essay. The essay should be seamless, treating the items in an integrated manner rather than separately and in isolation.

Exercise 6.32. Write an essay on solving cubic equations. Illustrate by computing solutions to

$$x^3 + 6x^2 + 3x + 18 = 0$$

and

$$x^3 - 3x^2 - 10x + 24 = 0.$$

Interweave a concrete treatment of the equations with the general discussion, including the following points.

(i) The reduction of arbitrary cubic equations to reduced cubic equations, so that Cardano's formula can be applied.

(ii) How Cardano's formula can be used to obtain not just one but three solutions to the reduced cubic equation. In particular, explain which cube roots to pair in the formula in order to obtain the three solutions.

(iii) How to compute cube roots of non-real complex numbers.

(iv) How to sidestep such computations by relying instead on trigonometric and inverse trigonometric functions.

Exercise 6.33. Write an essay on solving quartic equations. Illustrate by computing solutions to the quartic equations

$$x^4 - 4x^3 + 3x^2 + 8x - 10 = 0$$

and

$$x^4 - 42x^2 + 64x + 105 = 0.$$

Interweave a concrete treatment of the equations with the general discussion, including the following points.

(i) The reduction of arbitrary quartic equations to reduced quartic equations, and its application to the first quartic.

(ii) The solving of a reduced quartic equation, such as the one obtained, by writing the reduced quartic polynomial in it as a product of two quadratic polynomials with unknown coefficients, obtaining equations that the coefficients must satisfy, and reducing the solution of the equations to the problem of solving a cubic equation.

(iii) Solving the cubic equation in the example by guessing, using the solution to obtain a factorization of the quartic polynomial as a product of quadratic polynomials, obtaining solutions to the reduced quartic equation, and then finding solutions to the original quartic equation.

(iv) A description of the general form of the cubic equation to solve in studying a reduced quartic polynomial, illustrating this with the second example and obtaining

$$j^3 - 84j^2 + 1344j - 4096 = 0.$$

(v) The observation, after guessing a solution to this cubic equation, that we can use the procedure described for the first quartic to factor the second one as a product of two quadratic polynomials, followed by an explanation that we can instead describe the roots of the reduced quartic in terms of the three solutions of the cubic equation.

(vi) A list of eight candidates for solutions to the original quartic equation in terms of the three solutions to the cubic equation, and an explanation of which four to choose. (Provide an explanation that does not depend on substituting all eight candidates in the original equation.)

(vii) A description of a quartic equation whose solutions are the other four candidates, and an explanation of why they are solutions.

(viii) A discussion of how these ideas can be used to solve any quartic equation, once a method is available for solving cubic equations.

Exercise 6.34. Write an essay about the role the discriminant plays in understanding the solutions of quadratic, cubic, and quartic equations. Address the following issues.

(i) The definition of the discriminant of a polynomial of degree 2, 3, or 4 in terms of its roots, and the information it gives about the roots.

(ii) The formula for the discriminant of a polynomial of degree 2, 3, or 4 in terms of its coefficients, and how the formulas make it possible to obtain information about the roots from the coefficients.

(iii) The quadratic formula for the roots of a quadratic polynomial and the appearance of the discriminant within the formula.

(iv) Cardano's formula for the roots of a reduced cubic polynomial and the appearance of the discriminant within the formula.

(v) The way in which the sign of the discriminant of a cubic polynomial affects the calculations in Cardano's formula, with the need to compute cube roots of non-real complex numbers in certain cases.

(vi) How to combine information from the discriminant of a quartic polynomial with information from the quartic's coefficients to refine our understanding of the nature of the quartic's roots.

6.8 History

Lodovico Ferrari was the first person to develop a method for solving quartic equations. We already introduced him in Section 3.5, learning that he was Cardano's assistant, that Cardano described his quartic solution in *Ars Magna*, and that he participated in the famous dispute in 1548 with Tartaglia. Let us look at Cardano's presentation of Ferrari's work before moving on to contributions of two intellectual giants from the following centuries, René Descartes and Leonhard Euler.

Cardano presents the quartic solution in the penultimate chapter of *Ars Magna*, making the following transition [12, p. 237]:

> There is another rule, more noble than the preceding. It is Lodovico Ferrari's, who gave it to me on my request. Through it we have all the solutions for equations of the fourth power, square, first power, and number, or of the fourth power, cube, square, and number, and I set them out here in order.

What Cardano sets out are twenty families of quartic equations, depending on the signs of the coefficients.

After describing a procedure for handling the equations, Cardano works nine examples. The first example is introduced by [12, p. 239]:

> For example, divide 10 into three proportional parts, the product of the first and second of which is 6. This was proposed by Zuanne de Tonini de Coi, who said it would not be solved. I said it could, though I did not yet know the method [for doing so]. This was discovered by Ferrari.
>
> Let the mean be x. The first, then, will be $6/x$ and the third will be $x^3/6$. These equal 10. Multiplying all terms by $6x$, we will have
>
> $$60x = x^4 + 6x^2 + 36.$$

Cardano spends several pages on the details of this example's solution. It is the equation of Exercise 6.4. More examples follow, until he comes to

$$z^4 + 3 = 12z,$$

which we encountered in Exercise 6.5, and to an example of a non-reduced quartic. He concludes, "By this you know the methods for these rules if you have paid careful attention to the examples and the operations."

Let us turn to Descartes. He is one of the founders of both modern philosophy and modern mathematics. It is impossible to do justice to his life, his work, and his impact in a short space. We will have to be content with a brief survey of his contributions to algebra. But first, a few words about the broader context of his work and how algebra fit into it.

Descartes was born in 1596 in La Haye, France. In 1619 he had the now-famous dreams that inspired him to find a new basis for scientific inquiry. He moved to the Netherlands in 1628 and published *Discours de la méthode pour bien conduire sa raison et chercher la vérité dans les sciences*, or *A Discourse on the Method of Correctly Conducting One's Reason and Seeking Truth in the Sciences* [18], in 1637. Ian Maclean opens the introduction to his translation [18, p. vii] with the observation that it "marks one of the pivotal moments of Western European thought; it was the work of a formidably clever, radical, rigorous thinker, who in this short, informally presented introduction to his work threatened the very foundations of many prevailing philosophical beliefs, and set an agenda for enquiry into man and nature whose effects have lasted up to the present day."

The work consists of a preface and three essays. The preface, "A Discourse on the Method," is the most famous part, for it is here that Descartes seeks to "reform my own thoughts and build on a foundation which is mine alone." [18, p. 15]. Algebra is one of the areas to which he intends to apply his new foundation [18, p. 17]:

> As for ancient geometrical analysis and modern algebra, even apart from the fact that they deal only in highly abstract matters that seem to have no practical application, the former is so closely tied to the consideration of figures that it is unable to exercise the intellect without greatly tiring the imagination, while in the latter case one is so much a slave to certain rules and symbols that it has been turned into a confused and obscure art that bewilders the mind instead of being a form of knowledge that cultivates it. This was why I thought that another method had to be found.

In the next paragraph, he proposes his first precept, "never to accept anything as true that I did not *incontrovertibly* know to be so." This is a stiff challenge. Some pages later, in one of the most historic passages of western literature, Descartes arrives at his first item of knowledge [18, p. 28]:

> Because our senses sometimes deceive us, I decided to suppose that nothing was such as they lead us to imagine it to be. And because there are men who make mistakes in reasoning, even about the simplest elements of geometry, and commit logical fallacies, I judged that I was as prone to error as anyone else, and I rejected as false all the reasoning I had hitherto accepted as valid proof. Finally, considering that all the same thoughts which we have while awake can come to us while asleep without any one of them being true, I resolved to pretend that everything that had ever entered my head was no more true than the illusions of my dreams. But immediately afterwards I noted that, while I was trying to think of all things being false in this way, it was necessarily the case that I, who was thinking them, had to be something; and observing this truth: *I am thinking therefore I exist*, was so secure and certain that it could not be shaken by any of the most extravagant suppositions of the skeptics, I judged that I could accept it without scruple, as the first principle of the philosophy I was seeking.

This is Descartes' famed statement of certainty, "Cogito, ergo sum," traditionally translated as, "I think, therefore I am."

Having developed a method for obtaining knowledge from secure foundations, Descartes illustrates its effectiveness in three essays: *La Dioptrique* (treating optics), *Les Météores* (treating meteorology), and *La Géométrie* (treating, yes, geometry, but also algebra). Our interest is in *La Géométrie.* There is much more to be said about Descartes' philosophical work and his life, but let us leave it at that, adding only that in 1649, he moved from the Netherlands to Sweden to tutor Queen Christina, becoming ill a few months later and dying in 1650.

We turn to *La Géométrie*. (An English translation by David Eugene Smith and Marcia L. Latham with the title *The Geometry of René Descartes* was published in 1925, the translated pages interleaved with a facsimile of the original edition. It remains available in a reprinted version [19].) *La Géométrie* has three parts. One might not guess from the third part's title, "On the Construction of Solid or Supersolid Problems," that it is devoted to algebra. This makes sense once one learns that in Descartes' terminology, *solid* refers to equations of degree 3 and *supersolid* to equations of higher degree. This third part contains the first treatment of algebra written with language and algebraic notation that we would recognize. Let's survey some of its highlights.

In the fifth paragraph [19, p. 156], Descartes begins a sequence of "general statements ... concerning the nature of equations." The next sentence introduces a radical change in perspective: "An equation consists of several terms, some known and some unknown, some of which are together equal to the rest; or rather, all of which taken together are equal to nothing; for this is often the best form to consider."

What Descartes is proposing, and what we now take for granted, is that polynomial equations are best written in the form of a polynomial on one side of the equal sign and 0 on the other. Until this point equations would be written with the restriction that coefficients are positive and terms should be added, not subtracted, forcing the consideration of multiple forms of quadratic and cubic equations, as we have seen.

Another of Descartes' innovations is the use of letters near the end of the alphabet for variables and letters near the beginning or middle as constants, a notational convention that we take for granted. (Just before Descartes' birth, Viète had taken the first step in this direction, using vowels for variables and consonants for constants.)

In the ensuing pages, Descartes makes the first-ever statements of a series of now-standard results about polynomials, and in language we would find largely familiar. He speaks of the "dimension" of a polynomial or equation where we say "degree". He distinguishes between positive and negative roots of a polynomial, calling the positive ones "true," the negative ones "false." And, as we have seen, Descartes uses the word "nothing" in place of "zero." Given this terminology, the passage that comes after the sentence about setting polynomials equal to nothing is completely understandable [19, pp. 159–160]:

> Every equation can have as many distinct roots (values of the unknown quantity) as the number of dimensions of the unknown quantity in the equation.... It often happens, however, that some of the roots are false or less than nothing. ... It is evident from the above that the sum of an equation having several roots is always divisible by a binomial consisting of the unknown quantity diminished by the value of one of the true roots, or plus the value of one of the false roots. In this way, the dimension of an equation can be lowered. On the other hand, if the sum of the terms of an equation is not divisible by a binomial consisting of the unknown quantity plus or minus some other quantity, then this latter quantity is not a root of the equation.

We find here the first statements of two of our theorems: Theorem 1.6, that the number of roots of a polynomial is bounded by its degree, and Theorem 1.4, that if a is a root of the polynomial $f(x)$, then $x - a$ divides $f(x)$.

A few paragraphs later, Descartes explains how to pass from a polynomial equation with a given set of roots to a new one with roots shifted by a fixed quantity through a change of variable [19, pp. 163–164]:

> If the roots of an equation are unknown and it be desired to increase or diminish each of these roots by some known number, we must substitute for the unknown quantity throughout the equation, another quantity greater or less by the given number. Thus, if it be desired to increase by 3 the value of each root of the equation
>
> $$x^4 + 4x^3 - 19x^2 - 106x - 120 = 0$$
>
> put y in place of x, and let y exceed x by 3, so that $y - 3 = x$. Then for x^2 put the square of $y - 3$, or $y^2 - 6y + 9$; for x^3 put its cube, $y^3 - 9y^2 + 27y - 27$; and for x^4 put its fourth power, or $y^4 - 12y^3 + 54y^2 - 108y + 81$. Substituting these values in the above equation, and combining, we have ... $y^4 - 8y^3 - y^2 + 8y = 0$, or $y^3 - 8y^2 - y + 8 = 0$, whose true root is now 8 instead of 5 since it has been increased by 3.

New in this discussion is the abstract idea of transforming an equation with one set of roots to an equation with a different set of roots without knowing what the roots are. Descartes applies this procedure to pass from an arbitrary polynomial equation to a reduced one [19, p. 167]:

> Now this method of transforming the roots of an equation without determining their values yields two results which will prove useful: First, we can always remove the second term of an equation by diminishing its true roots by the known quantity of the second term divided by the number of dimensions of the first term, if these two terms have opposite signs; or, if they have the same signs, by increasing the roots by the same quantity.

Descartes works out some examples of this method, then explains how to transform a polynomial equation to a new one whose roots are all multiplied or divided by a given quantity. It is exciting to read these pages and see now-familiar algebraic notions and formulations coming to life for the first time.

Just a few pages later [19, pp. 180–183], before discussing cubic equations, Descartes turns to factoring quartic polynomials as products of quadratics. He passes to a reduced quartic (with p, q, and r as coefficients rather than our q, r, and s), then introduces the resolvent cubic without ado:

> Again, given an equation in which the unknown quantity has four dimensions ... we must increase or diminish the roots so as to remove the second term, in the way already explained, and then reduce it to another of the third degree, in the following manner: Instead of
>
> $$x^4 \pm px^2 \pm qx \pm r = 0$$
>
> write
>
> $$y^6 \pm 2py^4 + (p^2 \pm 4r)y^2 - q^2 = 0.$$
>
> For the ambiguous sign put $+2p$ in the second expression if $+p$ occurs in the first; but if $-p$ occurs in the first, write $-2p$ in the second ... . For example, given
>
> $$x^4 - 4x^2 - 8x + 35 = 0$$
>
> replace it by
>
> $$y^6 - 8y^4 - 124y^2 - 64 = 0.$$
>
> For since $p = -4$, we replace $2py^4$ by $-8y^4$; ... .
>
> Similarly, instead of
>
> $$x^4 - 17x^2 - 20x - 6 = 0$$
>
> we must write
>
> $$y^6 - 34y^4 + 313y^2 - 400 = 0,$$
>
> for 34 is twice 17, and 313 is the square of 17 increased by four times 6, and 400 is the square of 20.

We do not bother with the distinction Descartes maintains between positive and negative coefficients, and therefore would dispense with his clarifications on how to choose signs in writing the resolvent cubic equation.

Descartes presents a third example of a quartic and its resolvent cubic, one with coefficients that are algebraic expressions rather than specific numbers, then explains how to use the resolvent cubic to factor a quartic [19, p. 184]:

> If, however, the value of y^2 can be found, we can by means of it separate the preceding equation into two others, each of the second dimension, whose roots will be the same as those of the original equation. Instead of $x^4 \pm px^2 \pm qx \pm r = 0$, write the two equations
>
> $$+x^2 - yx + \frac{1}{2}y^2 \pm \frac{1}{2}p \pm \frac{q}{2y} = 0$$

> and
>
> $$+x^2 + yx + \frac{1}{2}y^2 \pm \frac{1}{2}p \pm \frac{q}{2y} = 0.$$
>
> ... It is then easy to determine all the roots of the proposed equation.

This is our Theorem 6.2.

Up to this point, Descartes has yet to discuss Cardano's formula. It has not been needed, as the values of y in his examples can be found directly and the quartics then factored as products of quadratics, as he illustrates in the sentences that immediately follow. Rather than continuing to read his discussion, let's do the work ourselves:

Exercise 6.35. Solve two quartic equations from *La Géométrie*. In both cases, a root of the the resolvent cubic can be found by guessing. (Try squares of small positive integers.)

(i) $x^4 - 4x^2 - 8x + 35 = 0.$

(ii) $x^4 - 17x^2 - 20x - 6 = 0.$

Descartes closes *La Géométrie* with a wish [19, p. 240]: "I hope that posterity will judge me kindly, not only as to the things which I have explained, but also as to those which I have intentionally omitted so as to leave to others the pleasure of discovery."

A little over a century later, Euler gave his own account of quartic equations in *Elements of Algebra* [26]. In Section 5.8 we discussed his treatment of Cardano's formula, which appears in Chapter XII of Section 4. The next three chapters treat the quartic. In Chapter XIII, Euler deals with some special families of quartics, then in Chapter XIV he presents the "rule of Bombelli," which is actually Ferrari's approach, misattributed by Euler. It is in Chapter XV, "Of a new method of resolving equations of the fourth degree," that Euler obtains the result that constitutes our Theorem 6.4.

Rather than using Descartes' approach, as given in Theorem 6.2, Euler hypothesizes from the start [26, p. 283] that a reduced quartic equation has a root that can be expressed as the sum of three square roots, then searches for an expression for the cubic polynomial whose roots these would be:

> We will suppose that the root of an equation of the *fourth* degree has the form, $x = \sqrt{p} + \sqrt{q} + \sqrt{r}$, in which the letters p, q, r express the roots of an equation of the *third* degree, such as $z^3 - fz^2 + gz - h = 0$; so that $p + q + r = f$; $pq + pr + qr = g$; and $pqr = h$.

We based our treatment of the quartic on Descartes' resolvent cubic, which led to the expression in Theorem 6.4 that is one-half the sum of three square roots. We could instead have absorbed those pesky 2s into the square root signs, obtaining three numbers that are one-fourth the roots of Descartes' resolvent. This would lead to a new resolvent cubic that is slightly different, and that is the cubic that Euler obtains.

Euler proceeds much as we did in Section 5.2 when we derived Cardano's formula for the roots of a reduced cubic polynomial from its resolvent quadratic. (See Theorems 5.2 and 5.3.) After a half-page of elementary calculations [26, p. 283], which can be done as an exercise, Euler arrives at the conclusion [26, p. 284]:

> Since, therefore, the equation
>
> $$x^4 - ax^2 - bx - c = 0,$$
>
> gives the values of the letters f, g, and h, so that
>
> $$f = \frac{1}{2}a, \; g = \frac{1}{16}a^2 + \frac{1}{4}c, \text{ and } h = \frac{1}{64}b^2, \text{ or } \sqrt{h} = \frac{1}{8}b,$$
>
> we form from these values the equation of the third degree $z^3 - fz^2 + gz - h = 0$, in order to obtain its roots by the known [Cardano's] rule. And if we suppose those roots, 1. $z = p$, 2. $z = q$, 3. $z = r$, one of the roots of our equation of the fourth degree must be, by the supposition,
>
> $$x = \sqrt{p} + \sqrt{q} + \sqrt{r}.$$
>
> This method appears at first to furnish only one root of the given equation; but if we consider that every sign $\surd$ may be taken negatively, as well as positively, we immediately perceive that this formula contains all the four roots.

A short discussion follows, like the last part of Exercise 6.14 and of Exercise 6.16. Euler explains that there are eight possibilities for the sum of the three square roots, but the correct four are chosen by using the rule that the square roots' product must equal $b/8$. Euler's notation doesn't match ours, nor does his resolvent cubic, but it is easy to check that with the appropriate change in the coefficient labels and signs and a change of variable by a factor of 4, his result coincides with Theorem 6.4.

Procedure in hand, Euler solves the equation $x^4 - 25x^2 + 60x - 36 = 0$ [26, pp. 284–285]. Let's do so too.

Exercise 6.36. Solve

$$x^4 - 25x^2 + 60x - 36 = 0$$

using the version of Euler's approach given in Theorem 6.4. One root of its resolvent cubic can be found by testing various integer squares. Using it, factor the cubic and find the other two roots. Euler constructed a very simple example indeed.

After that example, Euler demonstrates how a general quartic equation can be converted to a reduced quartic equation, thereby showing that his new rule is applicable to all quartics. Next comes the following observation [26, p. 286]:

> This is the greatest length to which we have yet arrived in the resolution of algebraic equations. All the pains that have been taken in order to resolve equations of the fifth degree, and those of higher dimensions, in the same manner, or, at least, to reduce them to inferior degrees, have been unsuccessful: so that we cannot give any general rule for finding the roots of equations, which exceed the fourth degree.

We will take up the story of the quintic next.

7

Higher-Degree Polynomials

We have devoted chapters to quadratic, cubic, and quartic polynomials. This pattern cannot continue through all degrees, and not just because the resulting book would be infinitely long. It turns out that results of the sort we have obtained do not exist for polynomials in degree greater than four. Therefore, we will content ourselves with a survey of some central results about higher-degree polynomials, combining proof sketches (or no proofs at all) with historical discussions. The chapter ends with a proof of the fundamental theorem of algebra.

7.1 Quintic Polynomials

We have obtained the quadratic formula, Cardano's formula, and Euler's formula for solutions of polynomial equations up through degree four. What about quintic polynomials? By Theorem 1.14, quintics have real roots. It is natural to seek a formula for a root like our other formulas, one involving the polynomial's coefficients and the operations of sum, product, quotient, and nth root extraction. If we had such an expression, we could solve quintic equations *by radicals*, the word "radical" referring to an nth root. Through the seventeenth and eighteenth centuries, progress was made in simplifying the problem of solving quintic equations, but no solution by radicals was found. Mathematicians began to suspect that a solution may not exist, and this was shown to be true in the nineteenth century. In this section, we will describe some of the highlights of the quintic's history, sketching results but providing no proofs.

Our first step in dealing with a cubic or quartic equation was to change variables to pass to a reduced equation, one in which the term of second highest degree drops out. This can be done for quintics as well. In fact, it can be done in general: Given

$$x^n + a_{n-1}x^{n-1} + \cdots + a_1 x + a_0 = 0,$$

we can replace x by $y = x - (a_{n-1}/n)$ to obtain

$$y^n + b_{n-2}y^{n-2} + \cdots + b_1 y + b_0 = 0,$$

which has no y^{n-1} term. It is called a *reduced* equation of degree n. The coefficients b_i are expressible in terms of the a_i. If we can solve the reduced equation, we can add a_{n-1}/n to each solution to obtain the solutions of the original equation.

The reduction of the problem of solving a quintic equation to that of solving a reduced quintic equation is straightforward, relying on an obvious linear change of variable. Ehrenfried Walter von Tschirnhausen (1651–1708) introduced the idea of using non-linear changes of variable to eliminate additional terms in an equation. He showed how to pass from a reduced equation of degree n to an equation of the form

$$z^n + c_{n-3}z^{n-3} + \cdots + c_1 z + c_0 = 0,$$

where the new coefficients are expressible in terms of the old ones. The equation has no terms of degrees $n-1$ or $n-2$ and is called a *principal* equation. If we can solve a principal equation, then Tschirnhausen's change of variable process allows us to pass back to solutions of the reduced equation, perhaps having to calculate some square roots to do so. From the solutions to the reduced equation, we get solutions of the original equation.

Applying Tschirnhausen's idea to a reduced equation of degree 3, we obtain a cubic equation of the form $z^3 + c = 0$, which can be solved by computing cube roots. Square root calculations lead to the roots of the reduced equation and adding a suitable constant leads to the roots of a general cubic. Cardano's formula can be derived in this way.

The process of passing from a reduced equation to a principal equation is an example of a general process called a *Tschirnhausen transformation*. Applying it to quintic polynomials, we are led to quintic equations of the form

$$z^5 + cz^2 + dz + e = 0.$$

Erland Bring (1736–1798) used more complicated Tschirnhausen transformations to go one step further. He showed in 1786 that the problem of solving a quintic equation can be reduced to that of solving a quintic equation of the form

$$w^5 + dw + e = 0.$$

Such equations are now called quintic equations in *Bring-Jerrard* form, in honor of Bring and of George Jerrard (1804–1863), who used the same idea to study equations of higher degree.

Several eighteenth-century mathematicians tried to solve the general quintic equation, including Bring, Etienne Bézout, Edward Waring, Lagrange, and—no surprise—Euler. We saw in Section 5.2 that Euler solved reduced cubic equations in *Elements of Algebra* [26] by determining, for a quadratic polynomial with roots U and V, the coefficients of a cubic polynomial whose roots have the form $\sqrt[3]{U} + \sqrt[3]{V}$. Going backwards, he associated to a reduced cubic polynomial its resolvent quadratic polynomial and found the roots of the cubic as sums of cube roots of the roots of the quadratic. He had a similar approach to solving reduced quartic equations using resolvent cubics, as we discussed in Section 6.8. We concluded that section with his remark that efforts (up to 1770) to resolve equations of the fifth degree had been unsuccessful.

Earlier, in a 1732 paper, Euler had conjectured that the roots of a quintic polynomial take the form

$$\sqrt[5]{A_1} + \sqrt[5]{A_2} + \sqrt[5]{A_3} + \sqrt[5]{A_4},$$

where A_1, A_2, A_3, and A_4 are roots of a quartic polynomial associated to the quintic. More generally, he conjectured such a result for polynomials of degree n, each root being the sum of nth roots of $n - 1$ quantities that would themselves be roots of an associated resolvent polynomial of degree $n - 1$. This would be a powerful extension of his approach to cubic and quartic polynomials. However, he was unable to make any progress on the conjecture, except for special families of polynomials.

The Italian mathematician Paolo Ruffini (1765–1822) was the first to put aside efforts to solve quintic equations and instead attempt to prove that there is no general solution by radicals. The title of his 1799 work says it all: *General theory of equations in which it is shown that the algebraic solution of the general equation of degree greater than 4 is impossible* [57]. In the introduction, Ruffini wrote [5, p.263]:

> The algebraic solution of general equations of degree greater than 4 is always impossible. Behold a very important theorem which I believe I am able to assert (if I do not err); to present the proof of it is the main reason for publishing this volume. The immortal Lagrange, with his sublime reflections, has provided the basis of my proof.

Ruffini sent Lagrange a copy of the book, but received no reply. In 1803, Ruffini published a paper with another proof. In 1806, he published yet another proof, and in 1813 he published a paper in which he expressed his disappointment at the poor reception his work received. As it turns out, Lagrange had read his work, but did not think the proof was complete and chose not to express his approval. Other contemporaries also did not find the proof complete or correct.

Raymond Ayoub has given an excellent appraisal of Ruffini's work in "Paolo Ruffini's Contributions to the Quintic" [5], along with an account of the work of Lagrange, Gauss, and others. It is clear in retrospect, as emerges from Ayoub's account, that Ruffini did not receive proper credit for his contributions. One exception was the response of the great French mathematician Augustin-Louis Cauchy, who wrote to Ruffini in 1821 [5, p. 271] that "your memoir on the general resolution of equations is a work which has always seemed to me worthy of the attention of mathematicians and which, in my judgement, proves completely the insolvability of the general equation of degree > 4."

It fell to the Norwegian mathematician Niels Abel (1802–1829) to provide the first widely recognized proof that there can be no general solution of quintic equations by radicals. His result was published in 1824 [1]. A few years later, the French mathematician Évariste Galois (1811–1832) provided another approach, one that would revolutionize algebra [31]. His proof can be found in every graduate-level text on algebra (such as *Abstract Algebra*, by David S. Dummit and Richard M. Foote [22, p. 629]) and some undergraduate texts as well, and his ideas continue to influence mathematics today.

The lives of Abel and Galois are of great interest. Both were mathematical geniuses of the first rank. Both died young. Galois is notorious for his dramatic death at 20 from wounds incurred in a duel, and for the image of him feverishly writing out his mathematical ideas in a letter on the eve of the duel. In both cases, we can only dream of the glorious discoveries they would have made had they lived longer, and reflect on the unfairness of life. Since their stories are well told elsewhere, we will move on. Especially recommended for further reading is Peter Pesic's *Abel's Proof: An Essay on the Sources and Meaning of Mathematical Unsolvability* [52].

There is one more twist to the tale. Ruffini deserves credit as the first mathematician to focus attention on proving that the quintic cannot be solved by radicals rather than searching for such a solution. But Gauss had similar inklings. In his 1799 doctoral dissertation, Gauss wrote [5, pp. 262–263]:

> After the labors of many geometers left little hope of ever arriving at the resolution of the general equation algebraically, it appears more

> and more likely that this resolution is impossible and contradictory.... Perhaps it will not be so difficult to prove, with all rigor, the impossibility for the fifth degree. I shall set forth my investigations of this at greater length in another place. Here it is enough to say that the general solution of equations understood in this sense [i.e., by radicals] is far from certain and this assumption [i.e., that any equation is solvable by radicals] has no validity at the present time.

Gauss never pursued the matter. Ayoub speculates that Gauss "did not attach very much importance to solvability by radicals," referring again to Gauss's dissertation, in which he wrote [5, p. 275]:

> what is called a solution to an equation is, in reality, nothing but the reduction of the equation to prime equations — the solution is not exhibited but symbolized — and if you express a root of the equation $x^n = H$ by $\sqrt[n]{H}$, you have not solved it nor done anything more than if you devise some symbol to denote a root of the equation $x^n + Ax^{n-1} + \cdots = 0$ and place the root equal to this symbol ... .

We made much the same point for quadratic equations at the end of Section 2.1, describing the quadratic formula as providing not a solution to a quadratic equation but a reduction of the solution to the problem of calculating square roots. Gauss dismisses the entire enterprise. His youthful views notwithstanding, the work of Abel and Galois has been justly celebrated.

The story of quintic equations does not end with Abel and Galois. As Gauss reminded us, a solution by radicals requires the extraction of nth roots, a process that is a problem of analysis, not algebra. Using standard functions of analysis, such as the exponential and logarithm, or cosine and inverse cosine, we can calculate nth roots of real and complex numbers. Perhaps by allowing an enlarged stock of functions from analysis, we can use algebra to reduce the solution of quintic equations to a form that can be handled. In the decades after the deaths of Abel and Galois, several mathematicians did this.

The first was Gotthold Eisenstein (1823–1852), in 1844. We have seen that the problem of solving the general quintic equation can be reduced by Tschirnhausen transformations to that of solving the Bring-Jerrard equation

$$w^5 + dw + e = 0.$$

One can reduce further to the problem of solving the quintic equation

$$v^5 + v - f = 0,$$

where f is a constant.

In a short note on solutions of equations of degree up to 4 [24], Eisenstein introduces functions ϕ and ψ that satisfy $\phi(\lambda)^2 = \lambda$ and $\psi(\lambda)^3 = \lambda$, observing that ϕ is defined up to multiples of ± 1 and ψ is defined up to multiples of the powers of ω. These are the square root and cube root functions. He expresses solutions of general cubic and quartic equations in terms of ϕ and ψ. He then concludes his paper with the remark that the roots of a general equation of degree 5 have a similar form, using in place of ϕ and ψ a function χ satisfying

$$\chi(\lambda)^5 + \chi(\lambda) = \lambda.$$

In a footnote, he offers a power series formula for χ:

$$\chi(\lambda) = \lambda - \lambda^5 + 10\frac{\lambda^9}{2!} - 15 \cdot 14\frac{\lambda^{13}}{3!} + 20 \cdot 19 \cdot 18\frac{\lambda^{17}}{4!} - \cdots.$$

We can think of χ as a generalized fifth-root function. Given a real number f, we can apply χ to obtain a new number $\chi(f)$ that is not quite a fifth-root of f: rather than satisfying $\chi(f)^5 = f$, it satisfies $\chi(f)^5 + \chi(f) = f$. But this means that $\chi(f)$ is a solution to our equation $v^5 + v = f$, and with such a solution available, we can solve any quintic equation.

Although quintic equations can't be solved by radicals, Eisenstein has demonstrated that we can come close. We simply need his power series, which lets us calculate a slight extension of fifth roots: the numbers v satisfying $v^5 + v = f$.

Papers by S.J. Patterson [51] and John Stillwell [62] contain illuminating discussions of Tschirnhausen transformations, the Bring-Jerrard form of the quintic, and Eisenstein's solution of the quintic. Stillwell comments [62, p 61] that Eisenstein added his solution to $v^5 + v = f$ "only as an afterthought," speculating that "Eisenstein may well have have considered this solution to be child's play, because it is based on a method he discovered for himself at age 14." (In another of Eisenstein's 1844 papers [25], he had described how a certain power series arose from his first mathematical research, in his fifteenth year.)

In 1858, Charles Hermite (1822–1901) used the Bring-Jerrard reduction to solve quintic equations, but in terms of a type of function known as a *modular function* [34]. Felix Klein (1849–1925) described another approach in his famous 1884 book *Lectures on the Icosahedron* [37].

A century later, Peter Doyle and Curt McMullen published a paper, "Solving the quintic by iteration" [21], in which they showed that quintic equations can be solved by a special iteration process involving "generally convergent" algorithms based on rational functions. (Rational functions are quotients of polynomial functions.) This is a result in the relatively new theory of complex dynamics, a subject of great contemporary interest. They also proved a modern analogue of the Abel-Galois unsolvability result, stating that polynomial equations of higher degree cannot be solved by their iteration process. In 1998, McMullen was one of four recipients of the Fields Medal at the International Congress of Mathematicians. The Fields Medal is the highest honor a mathematician can receive and is regarded as the equivalent in mathematics of the Nobel Prize.

7.2 The Fundamental Theorem of Algebra

The discovery of solutions to cubic equations in the sixteenth century led to the introduction of complex numbers, which in turn led to the realization that we can find solutions to all equations of degree four or less. What happens for higher degrees? That's the subject of this section.

The relevant mathematical result is known as the fundamental theorem of algebra. It has several formulations, which we will discuss without providing proofs. What we do discuss, in place of proofs, is the need to bring non-algebraic information to bear in order to find a proof. We then turn to the fundamental theorem's history, with special attention to a gap in some of the early proofs that can be filled through the introduction of the algebraic notion of a field. We will return to the fundamental theorem in Section 7.5, in order to provide a proof.

A quadratic equation has either two real roots or a pair of conjugate complex roots; a cubic equation always has a real root and either another two real roots or a pair of conjugate complex roots; a quartic equation has zero, two, or four real roots, any others occuring in conjugate complex pairs. For higher degrees, we know that every odd degree polynomial has a real root (Theorem 1.14), but we have no results for even degrees. Given our experience with polynomials of degrees 2 and 4, we might expect that even degree polynomials have roots provided we allow complex numbers as well as real numbers. The truth of this is the fundamental theorem of algebra.

Theorem 7.1 (fundamental theorem of algebra). *Any polynomial of positive degree with real coefficients has a root in the complex numbers.*

Theorem 4.2 states that if a complex number r is a root of a polynomial with real coefficients, then so is $\overline{r}$. We can use this to prove Theorem 7.2.

Theorem 7.2. *Let $f(x)$ be a positive-degree polynomial with real coefficients and let r a root of $f(x)$ that is a non-real complex number. Let*

$b = -(r + \overline{r})$ and $c = r\overline{r}$. Then b and c are real and $f(x)$ factors as

$$(x^2 + bx + c)g(x),$$

for a polynomial $g(x)$ with real coefficients.

Exercise 7.1. Prove Theorem 7.2.

Theorem 7.2 has the following consequence:

Theorem 7.3. *Let $f(x)$ be a polynomial of degree $n > 0$ with real coefficients.*

(i) $f(x)$ factors as a constant multiple of a polynomial of the form

$$(x - r_1)\dots(x - r_j)(x^2 + b_1x + c_1)\dots(x^2 + b_kx + c_k),$$

where j and k are non-negative integers satisfying $j + 2k = n$ and the coefficients occurring are real numbers.

(ii) The numbers $r_1, \dots, r_j$ are the real roots of $f(x)$, each value occurring as often as its multiplicity as a root.

(iii) The non-real roots of $f(x)$ are the roots of the quadratic factors $x^2 + b_ix + c_i$. There are k conjugate pairs of such roots, allowing repetitions.

The principal point of Theorem 7.3 is that any positive-degree polynomial with real coefficients factors as a product of linear and quadratic polynomials with real coefficients. This both follows from and implies Theorem 7.1, and is thus also called the fundamental theorem of algebra.

The fundamental theorem extends to polynomials with complex coefficients:

Theorem 7.4. *Let $f(x)$ be a polynomial of degree $n > 0$ with real or complex numbers as coefficients.*

(i) $f(x)$ has a root in the complex numbers. In fact, $f(x)$ has exactly n complex numbers as roots, counting multiplicities.

(ii) $f(x)$ factors as a constant multiple of a polynomial of the form

$$(x - r_1)\dots(x - r_n),$$

where $r_1, \dots, r_n$ are the complex numbers that are roots of $f(x)$.

Theorem 7.4 appears at first to be a significant extension of the fundamental theorem, but it follows almost immediately via a simple trick. Write $\overline{f}(x)$ for the polynomial obtained by conjugating each of the coefficients of $f(x)$. It's easily checked that the product $f(x)\overline{f}(x)$ is a polynomial with real coefficients. But then Theorem 7.3 can be applied to it, yielding Theorem 7.4.

Despite its name, the fundamental theorem cannot be proved by algebra alone. This should not be surprising. Even proving that the polynomial $x^2 - c$ has a real root for a positive real number c (Theorem 1.10) requires the intermediate value theorem, which is also needed to prove that every cubic polynomial with real coefficients has a real root (Theorem 1.14). We can hardly expect to prove the full fundamental theorem with anything less than the intermediate value theorem as a tool.

Typically, the more powerful the tools used from calculus or the broader field of analysis, the shorter the proof of the fundamental theorem. For example, the fundamental theorem can be proved in just two or three sentences [3, p. 122] from Liouville's theorem, which states that a bounded entire function must be a constant. (Joseph Liouville was a nineteenth-century French mathematician.) The hard work comes in proving Liouville's theorem. A proof also follows from the result from topology that a continuous real-valued function on a disk in the plane takes on a minimum and a maximum value, but a longer argument is needed. See for example the proof that Charles Fefferman published in 1967 while still an undergraduate [27].

From an algebraic perspective, the most attractive approach to proving the fundamental theorem would be to rely on the intermediate value theorem to prove that every positive real number has a real square root and every cubic polynomial with real coefficients has a real root, and then rely on algebra alone for the full theorem. A short, elegant proof follows from more advanced results of group theory and Galois theory, both of which grew from Galois's ideas in proving the unsolvability of quintic polynomials. See, for example, the exposition in [22, pp. 615–617]. Alternatively, we can give a proof using more elementary algebraic ideas. We will introduce these ideas in Section 7.4, then discuss the proof in Section 7.5. A useful reference is *The Fundamental Theorem of Algebra*, by Benjamin Fine and Gerhard Rosenberger [29], which presents proofs of the fundamental theorem using analysis, algebra, and topology.

Let's turn to some high points in the history of the fundamental theorem. Reinhold Remmert's article on the theorem provides an excellent overview [55, pp. 97–122]. See also Chapter 14 of John Stillwell's *Mathematics and its History* [61, p. 266] and Chapter 6 of Bashmakova and Smirnova's *The Beginnings and Evolution of Algebra* [8, pp. 98–99].

The first mathematician to state the theorem was Albert Girard, in 1629, in *Invention Nouvelle en l'Algèbre* [32]. We briefly examined Girard's account of cubic equations in Section 5.8. After discussing cubics, he states a theorem that begins, "All algebraic equations have as many solutions as the size of the highest quantity." Girard makes no effort to provide justification, but he illustrates the statement with examples. For the first one, Girard writes, "Given the equation $x^4 = 4x^3 + 7x^2 - 34x - 24$, the size of the highest quantity is 4, which signifies that there are 4 certain solutions, neither more nor less." Girard's final example is $x^4 = 4x - 3$, for which he lists the solutions $1, 1, -1 + \sqrt{-2}, -1 - \sqrt{-2}$. From this, we can infer that Girard intends his theorem to be understood in the context of complex numbers, with roots counted according to their multiplicities. Thus, he has provided a correct statement of the fundamental theorem.

Several of the leading mathematicians of the eighteenth century—Jean-le-Rond d'Alembert, Euler, Lagrange, and Pierre-Simon Laplace (1749–1827)—attempted to prove the fundamental theorem, but their arguments were not complete. (See Chapter 6 of Dunham's *Euler: The Master of Us All* [23] for an account of Euler's efforts.) One problem with these early proofs was their reliance on the implicit assumption that a polynomial has roots somewhere, where not being clear. Within this mysterious domain, they would then show that the roots actually lie in the complex numbers. As Remmert puts it [55, pp. 98–99], "until Gauss all mathematicians *believed* in the existence of solutions in some sort of no-man's land ... and tried imaginatively to show that these solutions were in fact complex numbers." It was Gauss who first pointed out this problem, in his doctoral dissertation of 1799 [55, p. 104]:

> Gauss begins his dissertation by a detailed critical examination of all previous attempts to prove the theorem known to him. This is not the place to discuss in detail the objections raised by the twenty-two year old student against the proofs of d'Alembert, Euler, and Lagrange—and thus against the leading mathematicians of the time ... . Gauss's main objection was that the existence of a point at which the polynomials take the value zero is always assumed and that this existence needs to be proved. Thus for example he reproaches Euler for using hypothetical roots.

Gauss was not yet aware of Laplace's proof. In 1815, he would subject Laplace to the same criticism [55, p. 105]: "The ingenious way in which Laplace dealt with this matter cannot be absolved from the main objections affecting all these attempted proofs."

Gauss offered not just criticism in his 1799 dissertation, but also a proof of his own, one that did not rely on the existence of roots in some unspecified domain. Rather, he set out to prove their existence from scratch, within the complex numbers. Although Gauss avoided the error of his predecessors, his proof, which is topological in nature, had gaps.

The first proof of the fundamental theorem that appears to be free of error is one given by Argand in 1814, relying on the existence of a minimum for a continuous function. He doesn't justify the existence of the minimum, something Cauchy would later do. Gauss would continue to give additional proofs using a variety of methods. His second proof, from 1816, is algebraic in nature, and correct. It completes Euler's argument.

Regarding the gap in the proofs of Gauss's predecessors, Remmert observes [55, p. 109]:

> Nowadays one can only speculate about how mathematicians before the beginning of the nineteenth century had visualized the solutions of equations in their mind's eye. It is difficult for us to understand why, until the time of Gauss, they had an unshakable belief in a kind of "extraterrestrial" existence of such solutions "somewhere or other," and then sought to show that these solutions were complex numbers.

With the development of algebra in the nineteenth century, it became a simple matter to construct the extraterrestrial solutions.

In studying polynomial equations, we typically wish to find solutions in some familiar domain of numbers, such as the rational numbers, the real numbers, or perhaps the complex numbers. As our historical discussion suggests, it may be convenient, at least provisionally, to search for solutions in a broader domain of numbers, one that may contain the complex numbers and be large enough to contain roots of the polynomial. If we can introduce such a domain, we can then work within it to show that the roots are in fact complex numbers. This leads to the algebraic notion of a *field*.

The families of numbers with which we are most familiar—integers, rationals, reals, and complexes—all have certain basic arithmetic properties in common. To describe them, let us write K to denote any of these four sets of numbers. Then K satisfies:

(1) *Commutativity*: Any two numbers a and b in K satisfy $a + b = b + a$ and $a \cdot b = b \cdot a$.

(2) *Associativity*: Any three numbers a, b, and c in K satisfy $(a+b)+c = a + (b + c)$ and $(a \cdot b) \cdot c = a \cdot (b \cdot c)$.

(3) *Distrtibutivity*: Any three numbers a, b, and c in K satisfy $a \cdot (b+c) = a \cdot b + a \cdot c$.

(4) *Additive Identity*: There is an *additive identity* in K, that is, a number o satisfying $o + a = a = a + o$ for any number a.

(5) *Additive Inverses*: For any number a in K, there is an *additive inverse*; that is, a number a' in K such that $a + a' = o$.

(6) *Multiplicative Identity*: There is a *multiplicative identity* in K, that is, a number ℓ satisfying $\ell \cdot a = a = a \cdot \ell$ for any number a.

There is a seventh property that K may satisfy:

(7) *Multiplicative Inverses*: For any number a in K other than the additive identity o, there is a *multiplicative inverse*; that is, a number a'' in K such that $a \cdot a'' = \ell$.

Any set K of numbers that satisfies properties (1) to (6) is called a *ring*. The integers, rationals, reals, and complexes all form rings, with 0 as the additive identity, $-a$ as the additive inverse of a number a, and 1 as the multiplicative identity. If property (7) is satisfied also, and if the additive and multiplicative identities of K are distinct, then K is a *field*. The integers aren't a field, since only 1 and -1 have multiplicative inverses. The rational, real, and complex numbers are fields. (The existence of multiplicative inverses for complex numbers was treated in Exercise 4.3.) It will be convenient to use standard notations for these fields: $\mathbb{Q}$ is the field of rational numbers, $\mathbb{R}$ is the field of real numbers, and $\mathbb{C}$ is the field of complex numbers. The letter 'Q' is chosen for the rational numbers since they are quotients of integers.

Rings and fields are treated in any standard undergraduate algebra course. We will not pursue the derivation of the basic results on them here. What matters for us is that a field is the natural setting in which to study polynomials and their roots.

We now ask, given a polynomial $f(x)$ whose coefficients are real numbers, if we can construct a field K containing, but possibly larger than, $\mathbb{C}$ that contains a root of $f(x)$. Once we have developed the basic results on rings and fields, it is not difficult to prove this and more:

Theorem 7.5. *Let $f(x)$ be a polynomial of positive degree n with real coefficients. There exist a field K containing $\mathbb{C}$ and elements $r_1, r_2, \ldots, r_n$ in K, repetitions allowed, such that*

$$f(x) = (x - r_1)(x - r_2) \cdots (x - r_n).$$

We call K a *field extension* of $\mathbb{C}$ and say that $f(x)$ *splits* into linear factors over K. We may also call K a *splitting field* for $f(x)$, although this term is usually reserved for a field of this type that is minimal in a suitable sense.

Theorem 7.5 is the result that the mathematicians of the eighteenth century were missing, or unjustifiably took for granted. It is purely algebraic, the proof requiring no property of the real numbers other than the fact that they form a field. Once it is available, the proofs of the fundamental theorem that Gauss criticized can now be completed [55, pp. 107–108]: "The Gaussian objection against the attempts of Euler-Lagrange and Laplace was invalidated as soon as Algebra was able to guarantee the existence of a splitting field for every polynomial. From that moment on, as Adolf Kneser already observed in 1888, the attempted proofs became in effect fully valid."

We will discuss Theorem 7.5 in Section 7.3 and use it in Section 7.5, where we present Laplace's 1795 proof of the fundamental theorem.

7.3 Polynomial Factorization

In studying roots of a polynomial $f(x)$, we have had occasion to look at factorizations of $f(x)$ as a product of lower-degree polynomials. For instance, our proof that a polynomial of degree n has at most n roots relied on Theorem 1.5, which states that a polynomial $f(x)$ with distinct roots $a_1, \dots, a_k$ can be factored as

$$(x - a_1)(x - a_2) \cdots (x - a_k)g(x)$$

for a polynomial $g(x)$. In this section, we will study factorizations of $f(x)$ more generally. The principal theme that emerges is that polynomial factorization is analogous to integer factorization. Moreover, a result on polynomials can often be proved by mimicking the proof of the corresponding integer result. A second theme is that the results do not require the polynomial coefficients to be real numbers. As we will see, the coefficients can come from any set satisfying the axioms of a field. Working with more general fields will allow us to sketch a proof of Theorem 7.5, the missing ingredient in some early proofs of the fundamental theorem of algebra.

Let's begin our study of polynomial factorization by determining the polynomial analogue to prime number. Given a non-zero polynomial $f(x)$ of degree n and a non-zero number c, we can factor $f(x)$ as the product of the degree 0 polynomial c and the degree n polynomial $c^{-1}f(x)$. This is as interesting as the factorization of an integer $n > 1$ as $n \cdot 1$. In both cases, we call such factorizations *trivial*. A factorization of n as a product rs of smaller positive integers is called *non-trivial*, as is a factorization of $f(x)$ as a product of lower-degree polynomials $g(x)h(x)$. An integer $n > 1$

is *prime* if it has no non-trivial factorizations. Equivalently, n cannot be factored as the product of two smaller positive integers. Similarly, we say that a positive-degree polynomial $f(x)$ is *irreducible* if it has no non-trivial factorizations; that is, it cannot be factored as the product of two lower-degree polynomials.

Any degree 1 polynomial is irreducible, since constants are the only lower-degree polynomials and no product of constants can have degree 1. An example of an irreducible polynomial of degree 2 is $x^2 + 1$. If it could be factored as a product of lower degree polynomials, then the resulting degree 1 factors would correspond to roots, but $x^2 + 1$ has no real roots. More generally, a quadratic polynomial $x^2 + bx + c$ either factors as a product $(x - r)(x - s)$, corresponding to real roots r and s, or has no such factorization. Which of the two cases occurs can be determined by the sign of the discriminant $b^2 - 4c$, allowing us to conclude that $x^2 + bx + c$ is irreducible precisely when $b^2 - 4c < 0$.

We are assuming that the numbers we allow as coefficients are real numbers. However, we may wish to select the allowable coefficients from other classes of numbers, and this will lead to different factorization results. For example, if we allow complex numbers as coefficients, then $x^2 + bx + c$ is never irreducible. If we allow only rational numbers as coefficients, then $x^2 - 2$ is irreducible, even though it factors as $(x - \sqrt{2})(x + \sqrt{2})$ when we allow real numbers as coefficients.

If we choose to work with real coefficients, then the fundamental theorem of algebra (in the form of Theorem 7.3) states that any polynomial of positive degree factors as a product of degree 1 and degree 2 polynomials, from which it follows that the irreducible polynomials are the degree 1 polynomials and the degree 2 polynomials with negative discriminant. Theorem 7.4 implies that when we allow complex numbers as coefficients, the irreducible polynomials are precisely the degree 1 polynomials. In contrast, if we allow only rational numbers as coefficients, there are irreducible polynomials of every positive degree. For instance, for any positive integer n, $x^n - 2$ is irreducible. These examples make it clear that the study of polynomial factorization will depend on our choice of allowable coefficients.

It turns out that there are many natural choices besides the rational numbers, the real numbers, and the complex numbers. What they have in common is that they are examples of fields.

We have described fields as collections of numbers with addition and multiplication rules that satisfy seven properties. A review of the properties reveals that nothing in their wording requires that what we are adding and multiplying be numbers. They can be any sort of object—polynomials,

M&M's, Lego bricks—as long as addition and multiplication rules are introduced that satisfy the properties. We will have a better perspective on the notion of a field if we introduce two examples besides rational, real, and complex numbers.

Binary arithmetic gives one example. Write $\mathbb{F}_2$ for the set of symbols 0 and 1 with the familiar addition rules $0 + 0 = 0$ and $0 + 1 = 1 + 0 = 1$ along with the not-so-familiar rule $1 + 1 = 0$. For multiplication, we will adopt the usual rules $0 \times 0 = 0 \times 1 = 1 \times 0 = 0$ and $1 \times 1 = 1$.

Exercise 7.2. Under the definitions for addition and multiplication, verify that $\mathbb{F}_2$ is a field, with 0 as the additive identity and 1 as the multiplicative identity.

We have constructed a field with just two elements. It is important to recognize that the elements, 0 and 1, are not the usual numbers zero and one shared by $\mathbb{Q}$, $\mathbb{R}$, and $\mathbb{C}$. Rather, they are symbols satisfying the addition and multiplication rules just introduced. We may wish to give them concrete meaning, but $\mathbb{F}_2$ as described is a field whether we do so or not.

One way to give meaning to 0 and 1 is to regard 0 as a symbol for the collection of all even integers and 1 as a symbol for the collection of all odd integers. The addition and multiplication rules for 0 and 1 then represent familiar arithmetic facts about the addition and multiplication of even and odd integers. For example, $0 + 1 = 1$ represents the fact that the sum of an even integer and an odd integer is odd, while $1 \times 1 = 1$ represents the fact that the product of two odd integers is again odd. But we needn't have this in mind when working with $\mathbb{F}_2$.

A second example of a field is the set $\mathbb{R}(t)$ of rational functions. A *rational function* is a fraction $f(t)/g(t)$, where $f(t)$ and $g(t)$ are polynomials with real coefficients and $g(t)$ is non-zero. Just as in describing rational numbers as the fractions formed from integers, we have to be precise, identifying two fractions $f(t)/g(t)$ and $h(t)/k(t)$ as the same if $f(t)k(t) = g(t)h(t)$. Addition and multiplication are performed in the usual way, using common denominators for addition. The fraction $f(t)/g(t)$ is 0 precisely when $f(t) = 0$. Assuming $f(t)/g(t)$ is non-zero, its multiplicative inverse is $g(t)/f(t)$.

Once a field K is chosen, we can study polynomials with coefficients in K. The standard notation for the set of polynomials is $K[x]$. Thus, $\mathbb{R}[x]$ consists of polynomials whose coefficients are real numbers—the polynomials with which we have worked for most of this book—and $\mathbb{F}_2[x]$ consists of polynomials whose coefficients lie in the field $\mathbb{F}_2$.

Given two polynomials $f(x)$ and $g(x)$ in $K[x]$, the addition and multiplication laws for K allow us to add and multiply $f(x)$ and $g(x)$ in the way defined in Section 1.2, obtaining new polynomials $f(x) + g(x)$ and $f(x)g(x)$ that also lie in $K[x]$. (Under these addition and multiplication rules, $K[x]$ satisfies the defining properties of a ring, the *polynomial ring* with coefficients in K.) For example, we can add and multiply polynomials whose coefficients lie in our new fields $\mathbb{F}_2$ and $\mathbb{R}(t)$.

Exercise 7.3. Let $x + 1$ and $x^2 + x + 1$ be polynomials in $\mathbb{F}_2[x]$. (That is, we regard the coefficients as elements of $\mathbb{F}_2$. Thus, 1 is not our usual number 1. It is the element 1 of $\mathbb{F}_2$.)

(i) Verify that $(x + 1) + (x^2 + x + 1) = x^2$.
(Hint: $1 + 1 = 0$.)

(ii) Verify that $(x + 1)(x^2 + x + 1) = x^3 + 1$.

Exercise 7.4. Let

$$2tx^2 - \frac{1}{t}x + 1 \quad \text{and} \quad \frac{3t^5 + 4t}{t^2 + 1}x^4 + (3t^2 - 5)x$$

be polynomials in $\mathbb{R}(t)[x]$. Compute their sum and product.

Given a field K, let's write 0 for its additive identity. In all our examples of fields besides $\mathbb{F}_2$, the additive identity is the actual number 0, but sometimes, as is the case with $\mathbb{F}_2$, we can't identify 0 with a number. It is simply a distinguished member of the field. We can define the notion of root for any polynomial $f(x)$ in $K[x]$: a *root* of $f(x)$ is an element a of K satisfying $f(a) = 0$. For example, the polynomial $x^2 + 1$ in $\mathbb{F}_2[x]$ has 1 as a root, since

$$1^2 + 1 = 1 + 1 = 0.$$

The definition of irreducible polynomial makes sense for polynomials with coefficients in any field: a polynomial $f(x)$ of positive degree in $K[x]$ is irreducible if it cannot be factored as the product of two polynomials of lower degree in $K[x]$. In the next exercise, we determine the irreducible polynomials of low degree in $\mathbb{F}_2[x]$.

Exercise 7.5. Consider polynomials in $\mathbb{F}_2[x]$.

(i) Show that x and $x + 1$ are the only polynomials of degree 1 in $\mathbb{F}_2[x]$, and that both are irreducible.

(ii) Show that x^2, $x^2 + x$, and $x^2 + 1$ are not irreducible. Show that $x^2 + x + 1$ is the only other polynomial of degree 2 and that it is irreducible.

(iii) Write all eight polynomials of degree 3 in $\mathbb{F}_2[x]$. Determine which are irreducible. For those that are not, factor them as products of irreducible polynomials of lower degree.

We can show, just as with $\mathbb{Q}[x]$, that for every positive integer n, there is an irreducible polynomial of degree n in $\mathbb{F}_2[x]$.

Once we broaden the study of polynomials to allow any field as the set of coefficients, we become interested in two types of theorems: those that are specific to the choice of K, and those that hold no matter what field K is chosen. The fundamental theorem of algebra is an example of the first type of theorem, as it is a statement about polynomials with real or complex coefficients. The results of Sections 1.2, 1.3, and 1.4 are examples of the second type. They make sense for polynomials with coefficients in any field K and the proofs work when real numbers are replaced by elements of K. For example, combining Theorems 1.3 and 1.4, and extending them to arbitrary fields, we obtain:

Theorem 7.6. *Let K be a field and let $f(x)$ be a polynomial in $K[x]$. If $x - a$ divides $f(x)$, then a is a root of $f(x)$. Conversely, if a is a root of $f(x)$, then $x - a$ divides $f(x)$.*

Also, Theorem 1.5, which we restated at the beginning of this section, holds for polynomials with coefficients in any field K:

Theorem 7.7. *Let K be a field and let $f(x)$ be a polynomial in $K[x]$. If $a_1, \dots, a_k$ are distinct roots of $f(x)$ in K, then there is a polynomial $g(x)$ in $K[x]$ such that*

$$f(x) = (x - a_1)(x - a_2) \cdots (x - a_k)g(x).$$

In the remainder of this section, we will look at results on polynomial factorization that, like these theorems, hold for any choice of coefficient field K. Let us begin with a refinement of Theorem 7.7. The multiplicity of an element a as a root of a polynomial $f(x)$ is the largest integer m for which $(x - a)^m$ divides $f(x)$.

Theorem 7.8. *Let K be a field and let $f(x)$ be a polynomial in $K[x]$ with exactly k distinct roots $a_1, \dots, a_k$ in K. For $1 \le i \le k$, let m_i denote the multiplicity of a_i as a root of $f(x)$. Then $f(x)$ has a factorization*

$$f(x) = (x - a_1)^{m_1}(x - a_2)^{m_2} \cdots (x - a_k)^{m_k} g(x)$$

for a polynomial $g(x)$ in $K[x]$ that itself has no roots in K.

We might try to prove Theorem 7.8 by extending the argument in the proof of Theorem 1.5. Certainly we can show by that argument that $f(x)$ factors as

$$(x-a_1)^{r_1}(x-a_2)^{r_2}\cdots(x-a_k)^{r_k}g(x)$$

for some positive integers $r_1, \ldots, r_k$ and for a polynomial $g(x)$ not divisible by any of the $x-a_i$. However, we must do a little more work to verify that each exponent r_i equals m_i. Let's sketch the argument, using Exercise 1.9. We will take $i = 1$, but the argument is the same for any index.

By the definition of m_1, there is a polynomial $h(x)$ not divisible by $x-a_1$ such that $f(x) = (x-a_1)^{m_1}h(x)$. Also by the definition of m_1, we must have $r_1 \leq m_1$. Setting our two factorizations of $f(x)$ equal to each other and canceling $(x-a_1)^{r_1}$, we obtain

$$(x-a_2)^{r_2}\cdots(x-a_k)^{r_k}g(x) = (x-a_1)^{m_1-r_1}h(x).$$

If $m_1 = r_1$, we are done. Otherwise, we find that $x-a_1$ divides the product on the left side. Making an argument like the one used in the proof of Theorem 1.6, we find that either $x-a_1$ divides one of the factors $x-a_i$ for $i > 1$, which is impossible, or it divides $g(x)$, which is also impossible. Therefore we have $m_1 = r_1$.

Recall the following fact about factorization of integers:

Theorem 7.9. *Any integer $n > 1$ can be factored as a product of prime numbers.*

It is easy to take this for granted. Yet, it requires proof. The proof is straightforward. If n is prime, we are done. Otherwise, by the definition of prime, n factors as a product rs for smaller positive integers r and s. If r or s is prime, we leave it as is. If not, we factor it as the product of smaller positive integers. We continue in this way as we test each new factor for primeness. Any non-prime factors become smaller with each round, so that after at most $n-1$ rounds, we will have factored n as a product of prime numbers.

Theorem 7.9 has a polynomial analogue, with essentially the same proof:

Theorem 7.10. *Let K be a field. Any polynomial $f(x)$ of positive degree in $K[x]$ can be factored in $K[x]$ as a product of irreducible polynomials.*

Let's sketch the argument. If $f(x)$ is itself irreducible, we are done. Otherwise, by the definition of irreducibility, $f(x)$ factors as a product $g(x)h(x)$ for lower-degree polynomials $g(x)$ and $h(x)$ in $K[x]$. If $g(x)$ or $h(x)$ is irreducible, we leave it as is. If not, we factor it as the product of lower-degree polynomials in $K[x]$. After at most n rounds, where n is the degree of $f(x)$, we will have factored $f(x)$ as a product of irreducible polynomials.

We can refine the statements of the last two theorems to obtain the next two.

Theorem 7.11. *Let n be an integer greater than 1. There exist prime numbers $p_1, \ldots, p_k$ and positive integers $m_1, \ldots, m_k$ satisfying*

$$n = p_1^{m_1} \cdots p_k^{m_k}.$$

Theorem 7.12. *Let K be a field and let $f(x)$ be a polynomial of positive degree in $K[x]$. Then there exist irreducible polynomials $p_1(x), \ldots, p_k(x)$ in $K[x]$ and positive integers $m_1, \ldots, m_k$ satisfying*

$$f(x) = p_1(x)^{m_1} \cdots p_k(x)^{m_k}.$$

The irreducible factors of $f(x)$ of degree 1 in the factorization of Theorem 7.12 correspond to roots, whereas irreducible factors of higher degree have no roots.

A counterpart to the existence of prime factorizations for integers is the famous theorem that such factorizations are unique. The formal statement of this requires some care. Here's one version:

Theorem 7.13. *Let n be an integer greater than 1. Suppose*

$$p_1^{m_1} \cdots p_k^{m_k}$$

and

$$q_1^{n_1} \cdots q_\ell^{n_\ell}$$

are two prime factorizations of n, for prime numbers $p_1 < p_2 < \cdots < p_k$ and $q_1 < q_2 < \cdots < q_\ell$ and for positive integer exponents m_i and n_j. Then $k = \ell$ and for each i between 1 and k, we have the equalities $p_i = q_i$ and $m_i = n_i$.

The key to proving Theorem 7.13 is a property of prime numbers that is analogous to the result we proved in Exercise 1.9 for polynomials of the form $x - a$: if a prime number p divides a product rs of positive integers, then p divides r or p divides s. This result goes back to Euclid. It is a consequence of the result known as the euclidean algorithm for finding the greatest common divisor of two positive integers. Given it, we can prove Theorem 7.13 in much the same way as Theorem 7.8. The difficulties are more ones of organization than of conception.

Euclid's key property of prime numbers has its counterpart for irreducible polynomials: given a field K, if an irreducible polynomial $p(x)$ in $K[x]$ divides the product $r(x)s(x)$ of two polynomials $r(x)$ and $s(x)$ in $K[x]$, then $p(x)$ divides $r(x)$ or $s(x)$. This is an extension of Exercise 1.9. We will not prove it. The proof is not difficult, paralleling as it does the proof of the integer result. We introduce the notion of greatest common divisor for polynomials, obtain a polynomial analogue of the euclidean algorithm, and prove the result, mimicking the classical integer arguments.

Once this property of irreducible polynomials is available, we can mimic the proof of Theorem 7.13 to obtain what is essentially an extension of Theorem 7.8, a unique factorization theorem for polynomials. The statement is more complicated than its integer counterpart because we can alter a factorization of a polynomial by inserting constant factors without changing it in an essential way, but this is the only issue: up to constant factors, the same irreducible polynomials occur in any factorization of a polynomial into irreducible polynomials with the same exponents.

With these general factorization theorems in place, we may next wish to study factorization problems with particular fields of coefficients, such as $\mathbb{Q}$ or $\mathbb{F}_2$. They turn out to be important, and difficult. Let us instead conclude by returning to the issue that prompted us to introduce the notion of a field in Section 7.2.

We saw that several eighteenth-century attempts to prove the fundamental theorem of algebra failed because of a missing ingredient, Theorem 7.5. This states, given a polynomial $f(x)$ in $\mathbb{R}[x]$ of positive degree n, that there exists a field K containing $\mathbb{C}$ and elements $r_1, r_2, \ldots, r_n$ in K satisfying

$$f(x) = (x - r_1)(x - r_2) \cdots (x - r_n).$$

The proof does not depend on working with the real numbers. It yields more generally, with no additional work, the following theorem.

Theorem 7.14. *Let K be a field and let $f(x)$ be a polynomial of positive degree in $K[x]$. There exists a field L containing K and elements $r_1, r_2, \ldots, r_n$ in L, repetitions allowed, such that*

$$f(x) = (x - r_1)(x - r_2) \cdots (x - r_n)$$

in $L[x]$.

The proof is not difficult, and can be found in many algebra texts (for example, [22, p. 536]). A full discussion would take us too far afield. Let us take a brief look at the essential issue, which is the following partial result:

Theorem 7.15. *Let K be a field and let $f(x)$ be a polynomial of positive degree in $K[x]$. There exists a field L containing K and a root r of $f(x)$ in L.*

If we can prove Theorem 7.15, we can iterate its construction to prove Theorem 7.14.

The construction of L is surprisingly easy. All we need to do is mimic the construction of $\mathbb{C}$ from $\mathbb{R}$, which was done to build a field larger than $\mathbb{R}$ that contains a root of $x^2 + 1$. In that construction, we created a new element, i, took $\mathbb{C}$ to be the set of polynomial expressions of degree at most one in i with real coefficients, and defined addition and multiplication in the obvious way, taking into account that i must satisfy $i^2+1 = 0$, or $i^2 = -1$.

For Theorem 7.15 we can assume $f(x)$ is monic. Suppose it has the form

$$x^n - a_{n-1}x^{n-1} - \cdots - a_1 x - a_0,$$

where we choose minus signs to simplify what follows. We introduce a new element r that will be the analogue for $f(x)$ of i for $x^2 + 1$. We want r to be a root of $f(x)$, which means r satisfies

$$r^n - a_{n-1}r^{n-1} - \cdots - a_1 r - a_0 = 0,$$

or

$$r^n = a_{n-1}r^{n-1} + \cdots + a_1 r + a_0.$$

We then take L to be the set of all polynomial-like expressions in r of degree at most $n - 1$ with coefficients from K.

Addition of expressions is defined in the obvious way. For multiplication, we first treat r as a polynomial indeterminate and multiply as usual. The result will be a polynomial in r, but it may have degree n or greater. We then use the last equation, which we can think of as a rewrite rule telling us how to replace r^n by a sum of lower-degree terms in r. Iterating, we can write any polynomial in r as a polynomial in r of degree less than n. Having defined addition and multiplication in L, we easily verify that L satisfies the defining conditions of a ring.

What requires closer examination is the additional requirement that fields must satisfy: that every non-zero element has a multiplicative inverse. Is this the case for L? It is not difficult to see that the answer is no whenever $f(x)$ fails to be irreducible. However, if $f(x)$ is irreducible, then L is a field. The proof of this, like the proof of unique factorization for polynomials, depends on polynomial analogues of integer results on greatest common divisors and the euclidean algorithm.

Theorem 7.15 follows. If $f(x)$ is irreducible, L is a field and we are done. Otherwise, we can replace $f(x)$ with one of its irreducible factors $g(x)$ in $K[x]$ and apply the same construction to $g(x)$. The result will be a field containing a root of $g(x)$ and hence a root of $f(x)$ as well.

The idea behind enlarging $\mathbb{R}$ in order to create a new field containing a square root of -1 is a powerful one. We see now that it can be used to expand any field to a larger one containing roots of a given polynomial.

7.4 Symmetric Polynomials

In our study of quadratic, cubic, and quartic polynomials, we found formulas for the discriminant in terms of the polynomial's coefficients. The discriminant can be defined for a polynomial of any degree, and the lower-degree examples suggest the possibility that it can again be written in terms of the coefficients. There are two issues. Does such an expression exist? If so, what is it? The answer to the first is yes, as we will see in this section. Finding the formula turns out to be a more difficult problem, which we will not pursue.

The theorem that the discriminant of a polynomial can be expressed in terms of the polynomial's coefficients is a special case of a far-reaching result on symmetric expressions in the roots of a polynomial. An example of a symmetric expression is $r^2s + rs^2$, which remains unchanged if we swap r and s. In contrast, $r^2s + 2rs^2$ changes under the swap and therefore is not symmetric. The general result—which will apply to discriminants—is that any symmetric expression in a polynomial's roots can be written in terms of the polynomial's coefficients. We will develop the relevant ideas in this section, and conclude with some history.

Let's begin by defining the discriminant. Suppose $f(x)$ is a polynomial of positive degree n, of the form

$$x^n - a_1x^{n-1} + a_2x^{n-2} - \cdots + (-1)^{n-1}a_{n-1}x + (-1)^n a_n.$$

We index the coefficients so that the subscript of a coefficient and the exponent of its associated power of x sum to n, and also we write the coefficients with alternating signs. The convenience of these choices will soon be clear. Assume that the coefficients are real numbers.

By the fundamental theorem of algebra, $f(x)$ factors as

$$(x - r_1)(x - r_2)\cdots(x - r_n),$$

where $r_1, \dots, r_n$ are the real or complex roots of $f(x)$, listed with possible repetitions. The *discriminant* Δ of $f(x)$ is defined to be the square of the product of differences of all the pairs of roots. In product notation,

$$\Delta = \prod_{1 \le i < j \le n} (r_i - r_j)^2.$$

We take the product over all roots, with repetitions, not over the set of distinct roots. If there is a repeated root, which means there is a pair of indices i and j with $i \neq j$ and $r_i = r_j$, then the difference $r_i - r_j$ is 0 and $\Delta = 0$. If no root is repeated, then the factors in the product are non-zero and $\Delta \neq 0$.

We are assuming that coefficients are real numbers. We could choose to work in the generality of Section 7.3, with an arbitrary field K as the field of coefficients for the polynomial $f(x)$. Theorem 7.14 ensures the existence of a larger field L that contains a set $r_1, \dots, r_n$ of roots for $f(x)$ for which $f(x)$ factors in $L[x]$ as

$$f(x) = (x - r_1)(x - r_2) \cdots (x - r_n).$$

We can then define the discriminant Δ of $f(x)$ as the product

$$\Delta = \prod_{1 \le i < j \le n} (r_i - r_j)^2.$$

However, we would have to prove that Δ doesn't depend on the field L and roots r_i. This is a journey we choose not to take.

For polynomials $f(x)$ of degree 2, 3, or 4, we found formulas for the discriminant that express it as a sum of integer multiples of products of the coefficients. For example, we found in Theorem 5.6 that the discriminant of $x^3 + bx^2 + cx + d$ is

$$18bcd - 4b^3d + b^2c^2 - 4c^3 - 27d^2.$$

We wish to show for $f(x)$ of any degree n that a formula like this exists: the discriminant of $f(x)$ can be written as a polynomial expression in the coefficients of $f(x)$ with integer coefficients.

The discriminant is not the only expression in the roots of $f(x)$ that we have written as a polynomial expression in its coefficients. In Theorem 5.14 we saw for a cubic polynomial $x^3 - a_1x^2 + a_2x - a_3$ with roots r_1, r_2, and r_3 that

$$\begin{aligned} a_1 &= r_1 + r_2 + r_3, \\ a_2 &= r_1r_2 + r_1r_3 + r_2r_3, \\ a_3 &= r_1r_2r_3. \end{aligned}$$

The proof amounts to factoring the cubic as $(x-r_1)(x-r_2)(x-r_3)$ and carrying out the multiplication of the degree-one terms. The simpler quadratic analogue was obtained for real roots in Exercise 2.5 and in general in Exercise 4.8, and the quartic analogue was Theorem 6.11.

A similar result holds for polynomials of higher degree. We have factored a general degree n polynomial $f(x)$ as

$$(x - r_1)(x - r_2)\cdots(x - r_{n-1})(x - r_n).$$

When we carry out the multiplication of the n terms, the result is the sum of the expressions we obtain by choosing from each term $x - r_i$ either the x or the $-r_i$ and then multiplying them together. For an integer k between 1 and n, the coefficient of x^{n-k} will be the sum of the terms we obtain when we choose x from $n - k$ of the factors and choose $-r_i$'s from k of the factors. We might for instance choose constants from the factors with indices $i_1, i_2, \ldots, i_k$. This would yield

$$(-1)r_{i_1} \cdot (-1)r_{i_2} \cdots (-1)r_{i_k} \cdot x^{n-k},$$

or

$$(-1)^k r_{i_1} r_{i_2} \cdots r_{i_k} x^{n-k}.$$

To get the entire coefficient of x^{n-k}, which we have written as $(-1)^k a_k$, we need to add all these expressions together. The result is

$$a_k = \sum_{1 \le i_1 < i_2 < \cdots < i_k \le n} r_{i_1} r_{i_2} \cdots r_{i_k}.$$

The notation is formidable, but what it says is that a_k is the sum of all products of k of the roots r_i. This gives us a sequence of polynomial expressions in the roots that equal the simplest of polynomial expressions in the coefficients: the coefficients themselves, up to sign. Two special cases are

$$a_1 = r_1 + r_2 + \cdots + r_n$$

and

$$a_n = r_1 r_2 \cdots r_n.$$

We have now introduced several polynomial expressions in the roots of a polynomial $f(x)$ that can be written in terms of its coefficients: the discriminant and the sums of products of roots. What connects them? Each remains unchanged if we reorder the roots. The roots are specific complex numbers, but we have assigned names to them—r_1, r_2, and so on—that do not take into account which is which. Our naming is arbitrary. When we sum them, or sum their two-by-two products, or sum their fifth powers, we are forming quantities that are independent of our choice of names. This is the case as well for the discriminant, a point that is worth a moment of thought.

By definition,

$$\Delta(r_1, \dots, r_n) = \prod_{1 \le i < j \le n} (r_i - r_j)^2.$$

If we reorder the roots r_i in some way, the order of the factors in the product will change, but the list of factors is unchanged: we choose every possible pair of roots, take its difference, square it, and multiply. Since multiplication doesn't depend on the ordering, the product that defines $\Delta(r_1, \dots, r_n)$ doesn't either. If we were to omit the squares in the product, the resulting product

$$\prod_{1 \le i < j \le n} (r_i - r_j)$$

would fail to be independent of the ordering of the r_i's. For example, consider what happens if we switch the order of r_1 and r_2 but leave the order of $r_3, \dots, r_n$ intact. The factor $r_1 - r_2$ changes to its opposite, $r_2 - r_1$. For any $i > 2$, the factors $r_1 - r_i$ and $r_2 - r_i$ switch with each other, but their product is unchanged. The remaining factors $r_i - r_j$ with $i, j > 2$ remain unchanged. The product as a whole, then, is changed to its opposite.

Expressions in the roots of $f(x)$ that remain unchanged under re-naming of the roots, such as the examples we have just examined, are called *symmetric*.

To proceed, we pass to a more abstract framework. The formulas we have obtained relating roots and coefficients of a polynomial make sense whatever the values of the coefficients are. In effect, we can regard them as variables, to be replaced by particular real numbers when we identify a specific polynomial of interest. If we are studying the polynomial $x^3 + 3x^2 - 2x + 7$, for example, then we substitute -3 for a_1, -2 for a_2 and -7 for a_3 (keeping our sign convention in mind).

This suggests the path we should take, which is to start over, working with a generic polynomial of degree n whose coefficients are variables or indeterminates. We can in this way discuss all degree n polynomials at once, specializing the variable coefficients to actual ones when we have a specific polynomial in mind.

Let us begin anew, then. Not only should the coefficients of a degree n polynomial be generic, but the roots should be too. In other words, we replace the roots r_i by variables. Fix a positive integer n and introduce the new variables or indeterminates $t_1, t_2, \dots, t_n$, which we think of as stand-ins for the roots of a degree n polynomial. We anticipate that when we study a specific polynomial, we will substitute the complex numbers that arise as roots of the polynomial for the t_is.

We are interested in polynomial expressions in the t_is with integer coefficients. The most general such expression can be written as

$$\sum a_{i_1,i_2,\ldots,i_n} t_1^{i_1} t_2^{i_2} \cdots t_n^{i_n}.$$

It is understood that the sum is finite and the coefficients $a_{i_1,i_2,\ldots,i_n}$ are integers, all but finitely many of them being 0. Just as the t_is are stand-ins for the roots of a given degree n polynomial, the expressions are stand-ins for polynomial expressions in the roots.

We already know some expressions that will interest us. For each k between 1 and n, let's define $e_k(t_1, \ldots, t_n)$ by

$$e_k(t_1, \ldots, t_n) = \sum_{1 \le i_1 < i_2 < \cdots < i_k \le n} t_{i_1} t_{i_2} \cdots t_{i_k}$$

and call it the kth *elementary symmetric polynomial*. Repeating the analysis we made for sums of products of roots, we obtain:

Theorem 7.16. *Let n be a positive integer and let $t_1, t_2, \ldots, t_n$ be n indeterminates. Then*

$$(x - t_1)(x - t_2) \cdots (x - t_{n-1})(x - t_n) = x^n + \sum_{k=1}^{n} (-1)^k e_k(t_1, \ldots, t_n) x^{n-k}$$

We also introduce, for each positive integer k, the *power sum polynomials* $p_k(t_1, \ldots, t_n)$, defined by

$$p_k(t_1, \ldots, t_n) = t_1^k + \cdots + t_n^k,$$

with $p_0(t_1, \ldots, t_n) = 1$. For $n \ge 2$, we introduce the polynomial $\Delta(t_1, \ldots, t_n)$ defined by

$$\Delta(t_1, \ldots, t_n) = \prod_{1 \le i < j \le n} (t_i - t_j)^2.$$

We call this the *discriminant*.

Let's return for a moment to the polynomial $f(x)$ with which we worked earlier in this section. It had the form

$$f(x) = x^n - a_1 x^{n-1} + a_2 x^{n-2} - \cdots + (-1)^{n-1} a_{n-1} + (-1)^n a_n.$$

with roots $r_1, \ldots, r_n$ and factorization

$$(x - r_1)(x - r_2) \cdots (x - r_n).$$

If, for each index i from 1 to n, we substitute r_i for t_i, then we pass from the generic polynomial

$$x^n + \sum_{k=1}^{n} (-1)^k e_k(t_1, \ldots, t_n) x^{n-k}$$

to the specific polynomial

$$f(x) = x^n - a_1 x^{n-1} + a_2 x^{n-2} - \cdots + (-1)^{n-1} a_{n-1} + (-1)^n a_n.$$

The coefficients a_i, which are real numbers, can be expressed in terms of the roots r_j:

$$a_i = e_i(r_1, \ldots, r_n).$$

Similarly, when we specialize the variables t_i to be the roots r_i of $f(x)$, the resulting number

$$\Delta(r_1, \ldots, r_n)$$

is the discriminant of $f(x)$.

We have observed that certain expressions in the roots of a polynomial, such as its discriminant, remain unchanged when we reorder the roots. We can now study such expressions.

We say a polynomial $p(t_1, \ldots, t_n)$ in the variables t_i with integer coefficients is *symmetric* if any reordering of the variables leaves the polynomial unchanged. We can introduce notation to make this more precise. We define a *permutation* of the set of numbers $1, \ldots, n$ to be a function π that takes the set to itself so that any two numbers $i \neq j$ are moved to two numbers $\pi(i)$ and $\pi(j)$ that remain unequal: $\pi(i) \neq \pi(j)$. With that notation, $p(t_1, \ldots, t_n)$ is symmetric if for every permutation π,

$$p(t_{\pi(1)}, \ldots, t_{\pi(n)}) = p(t_1, \ldots, t_n).$$

Our earlier observations made for a specific polynomial's roots carry over: the elementary symmetric polynomials, power sum polynomials, and discriminant are all examples of symmetric polynomials. In contrast, the square root of the discriminant (obtained by omitting the squares in the definition of the discriminant) fails to be symmetric, as will just about any polynomial expression in the t_is with integer coefficients that we write down at random.

It is easy to see that integer multiples, sums, and products of symmetric polynomials are symmetric. We can form sums of integer multiples of products of the elementary symmetric polynomials and obtain new symmetric polynomials. For example, with $n = 2$, since $e_1(t_1, t_2)$ and $e_2(t_1, t_2)$ are symmetric, so is

$$e_1(t_1, t_2)^2 - 4e_2(t_1, t_2).$$

Expanding, we find that

$$e_1(t_1, t_2)^2 - 4e_2(t_1, t_2) = (t_1 + t_2)^2 - 4t_1 t_2 = (t_1 - t_2)^2 = \Delta(t_1, t_2).$$

Thus, we have written the discriminant $\Delta(t_1, t_2)$ as a polynomial expression in elementary symmetric polynomials with integer coefficients.

This expression is familiar. What's new is the context we have now established in which we can ask whether such expressions exist for any n. That is, for a positive integer n, is $\Delta(t_1, \ldots, t_n)$ equal to a polynomial expression in the elementary symmetric polynomials $e_i(t_1, \ldots, t_n)$ with integer coefficients? If such an expression exists, then when we specialize the t_is to the roots r_i of a polynomial $f(x)$, the elementary symmetric polynomials specialize to the values $e_i(r_1, \ldots, r_n)$, which are up to sign the coefficients of $f(x)$.

Our main result is that every symmetric polynomial, not just the discriminant, can be obtained as a polynomial expression in the elementary symmetric polynomials. This is the fundamental theorem on symmetric polynomials.

Theorem 7.17. *Let n be a positive integer. Every symmetric polynomial in the variables $t_1, \ldots, t_n$ with integer coefficients is an integer-coefficient polynomial expression in the elementary symmetric polynomials*

$$e_1(t_1, \ldots, t_n), \ldots, e_n(t_1, \ldots, t_n).$$

That is, given a symmetric polynomial $p(t_1, \ldots, t_n)$, there is an n-variable integer-coefficient polynomial q for which

$$p(t_1, \ldots, t_n) = q(e_1(t_1, \ldots, t_n), \ldots, e_n(t_1, \ldots, t_n)).$$

As a special case of Theorem 7.17, we obtain:

Theorem 7.18. *Let n be an integer, $n \geq 2$. The discriminant*

$$\Delta(t_1, \ldots, t_n) = \prod_{1 \leq i < j \leq n} (t_i - t_j)^2$$

is a polynomial expression in the elementary symmetric polynomials with integer coefficients.

We obtain the same result for power sum polynomials, too, but we can say more, as we will later in the section.

To see the power of Theorem 7.18, let's return to our standard degree n polynomial

$$x^n - a_1 x^{n-1} + a_2 x^{n-2} - \cdots + (-1)^{n-1} a_{n-1} + (-1)^n a_n$$

with roots $r_1, \ldots, r_n$. Suppose we are able to find the explicit n-variable polynomial q for which

$$\Delta(t_1, \ldots, t_n) = q(e_1(t_1, \ldots, t_n), \ldots, e_n(t_1, \ldots, t_n)).$$

We know that when we substitute the roots r_i for the variables t_i, the elementary symmetric polynomials e_k specialize to the coefficients a_k: $e_k(r_1, \ldots r_n) = a_k$. We can therefore use $\Delta(t_1, \ldots, t_n)$ to calculate the discriminant $\Delta(r_1, \ldots, r_n)$ of $f(x)$ in terms of the coefficients a_k:

$$\begin{aligned} \Delta(r_1, \ldots, r_n) &= q(e_1(r_1, \ldots, r_n), \ldots, e_n(r_1, \ldots, r_n)) \\ &= q(a_1, a_2, \ldots, a_n). \end{aligned}$$

We have obtained expressions for $\Delta(t_1, \ldots, t_n)$ in terms of the elementary symmetric polynomials for $n = 2, 3, 4$. Theorem 7.18 ensures that there is such an expression for any n. They can be quite complicated. For example, the discriminant of the principal quintic polynomial

$$x^5 + cx^2 + dx + e,$$

is given by

$$\Delta = 108c^5 e - 27c^4 d^2 + 2250c^2 de^2 - 1600cd^3 e + 256d^5 + 3125e^4.$$

The formula for the discriminant of a general quintic polynomial is a sum of 59 terms! As n increases, it becomes increasingly difficult to determine the expression for the discriminant.

The proof of Theorem 7.17 is not difficult conceptually, but requires a good choice of notation. Let's see how it is proved for $n = 2$.

Since we have only two variables, we will dispense with indices. Let's write the variables as t and u. The elementary symmetric polynomials are $t + u$ and tu. What we are to prove is that any symmetric polynomial in t and u is a polynomial expression in $t + u$ and tu.

Suppose $p(t, u)$ is a symmetric polynomial. Since there are only two variables t and u, being symmetric means that if we switch t and u in the expression for $p(t, u)$, we leave the polynomial unchanged. In particular, if

$t^m u^n$ occurs in $p(t, u)$ with a non-zero coefficient a, then $t^n u^m$ must also occur in $p(t, u)$ with coefficient a. It follows that $p(t, u)$ must be a sum of integer multiples of terms of the form

$$t^m u^n + t^n u^m,$$

for non-negative integers m and n with $m < n$, and integer multiples of terms of the form

$$t^m u^m,$$

for non-negative integers m. Therefore, if we show that these expressions can be written as polynomials in $t + u$ and tu, it will follow that every symmetric polynomial in t and u can be written in terms of $t + u$ and tu.

We can simplify further. Given non-negative integers $m < n$, let $n = m + k$. Then

$$t^m u^n + t^n u^m = t^m u^m (t^k + u^k) = (tu)^m (t^k + u^k).$$

Thus we can combine the two cases into one, and it suffices to show for all non-negative integers m and k that

$$(tu)^m (t^k + u^k)$$

is a polynomial expression in $t + u$ and tu. Since $(tu)^m$ is already such an expression, we need only show for every non-negative integer k that the sum $t^k + u^k$ is a polynomial expression in $t + u$ and tu.

This is easily done. If $k = 0$, the sum $t^0 + u^0$ is 1, and if $k = 1$, it is $t + u$. These are both polynomial expressions in $t + u$. Suppose we already know for an integer $k > 1$ that $t^{k-2} + u^{k-2}$ and $t^{k-1} + u^{k-1}$ are polynomial expressions in $t + u$ and tu. Then

$$t^k + u^k = (t^{k-1} + u^{k-1})(t + u) - tu(t^{k-2} + u^{k-2})$$

shows that $t^k + u^k$ is also a polynomial expression in $t + u$ and tu. In this way, starting from the knowledge that our result holds for $k = 0$ and $k = 1$, we can work our way through all possible positive integers k, concluding for any positive integer k that the sum $t^k + u^k$ is a polynomial expression in $t + u$ and tu. (We are implicitly using the principle of mathematical induction, which will be discussed in Section 7.5.) By the earlier reductions, we have proved that every symmetric polynomial in t and u is a polynomial expression in $t + u$ and tu. This proves Theorem 7.17 for polynomials in two variables.

Theorem 7.17 is proved for polynomials in more variables by a similar bootstrap process. The organization of the proof is more complicated, but the idea is the same. See [22, pp. 621–622] for an outline.

In proving Theorem 7.17 for two variables t and u, we were able to reduce to the case of the power sum polynomials $t^k + u^k$, which we then proved to be polynomial expressions in $t + u$ and tu. For general n, Theorem 7.17 guarantees that the power sum polynomials $p_k(t_1, \ldots, t_n)$ can be expressed as polynomials in $e_1(t_1, \ldots, t_n), \ldots, e_n(t_1, \ldots, t_n)$ with integer coefficients. But we can do better, obtaining an explicit formula that relates power sum polynomials and elementary symmetric polynomials. In this formula, Theorem 7.19, it is understood that $e_0(t_1, \ldots, t_n) = 1$ and $e_k(t_1, \ldots, t_n) = 0$ for $k > n$. We omit the proof. (See [22, pp. 618–619] for a discussion.)

Theorem 7.19 (Girard-Newton identities). *Let k and n be positive integers. Then*

$$ke_k(t_1, \ldots, t_n) = \sum_{i=1}^{k} (-1)^{i-1} e_{k-i}(t_1, \ldots, t_n) p_i(t_1, \ldots, t_n).$$

By using $e_0(t_1, \ldots, t_n) = 1$ and moving terms around, we can re-write the kth Girard-Newton identity as

$$\begin{aligned} p_k(t_1, \ldots, t_n) = {} & (-1)^{k-1} ke_k(t_1, \ldots, t_n) \\ & - \sum_{i=1}^{k-1} (-1)^{k-i} e_{k-i}(t_1, \ldots, t_n) p_i(t_1, \ldots, t_n). \end{aligned}$$

When $k = 1$, this is

$$e_1(t_1, \ldots, t_n) = p_1(t_1, \ldots, t_n),$$

which holds since both sides are equal to $t_1 + \cdots + t_n$ by definition. From this starting point, as k increases, the other Girard-Newton identities allow us to express each power sum $p_k(t_1, \ldots, t_n)$ recursively as a polynomial in the elementary symmetric polynomials. When we specialize to a degree n polynomial

$$x^n - a_1 x^{n-1} + a_2 x^{n-2} - \cdots + (-1)^{n-1} a_{n-1} + (-1)^n a_n,$$

we are able to write the power sums $p_k(r_1, \ldots, r_k)$ in terms of the coefficients a_i. Girard wrote such formulas explicitly for $k \leq 4$. Let's conclude the section with a look at his work.

Prior to Girard, Viète obtained formulas in special cases for the coefficients of a polynomial as sums of products of the roots. Girard was the first to state such formulas in general. In his 1629 book *Invention Nouvelle en l'Algèbre* [32], Girard stated the theorem that any polynomial of degree n has n roots. Only with the presence of n roots can we even contemplate expressions for the coefficients of the polynomial in terms of the roots. Girard provided them in the second part of the same theorem.

Before stating the theorem, Girard introduced a notion he called a *faction*:

> When several numbers are proposed, their sum is called the first faction; the sum of all the products two-by-two is called the second faction; the sum of all the products three-by-three is called the third faction; and so on up to the end, with the product of all the numbers called the last faction. There are as many factions as the numbers proposed.

The second part of the theorem then states:

> The first faction of the solutions [of a polynomial equation] equals the first coefficient, the second faction of the same is equal to the second coefficient, the third the third, and so on, so that the last faction is equal to the last, and this according to the signs, which one observes are in alternating order.

Girard offers no proof. But the proof is clear, provided we understand a polynomial of degree n to have n roots, as Girard states, and provided we understand that the polynomial factors as the product of terms $x - r$, as r runs through the n roots, which he surely did.

A few pages later, Girard introduces the expressions in the roots r_i that we call the power sum polynomials. He then writes formulas for them in terms of the coefficients a_i, for powers up to 4:

$$\begin{aligned} r_1 + \cdots + r_n &= a_1, \\ r_1^2 + \cdots + r_n^2 &= a_1^2 - 2a_2, \\ r_1^3 + \cdots + r_n^3 &= a_1^3 - 3a_1a_2 + 3a_3, \\ r_1^4 + \cdots + r_n^4 &= a_1^4 - 4a_1^2a_2 + 4a_1a_3 + 2a_2^2 - 4a_4. \end{aligned}$$

These are the first four Girard-Newton identities of Theorem 7.19. Girard doesn't continue with higher exponents, but it is clear that a sequence of such formulas exists.

Isaac Newton (1643–1727) obtained the same result in 1707 [47, pp. 392–393], again without proof. (In supplementary notes immediately after Newton's statement of the result, Reverend Wilder provides a proof [47, pp.

393–396].) The identities are often named after Newton alone, but credit is due to both.

Girard's observation that certain symmetric expressions in the roots of a polynomial can be written in terms of the polynomial's coefficients comes in for praise in H. Gray Funkhouser's 1930 paper, "A short account of the history of symmetric functions of roots of equations" [30, pp. 358, 360]:

> As is the case with the development of algebra as a whole, the subject of symmetric functions to a perhaps greater extent waited upon the introduction and improvement of symbolism. It was hard to perceive with any clarity relations between roots and coefficients when the equation was wrapped up in a paragraph of words representing it. Prior to the time of Franciscus Vieta, who was among the first to employ letters to represent numbers, we find very little trace of the thing we are seeking. ...
>
> We can appreciate how great an advance Vieta made when there is considered the few fragmentary statements that have come from his predecessors. He paves the way and provides an introduction to the first man who really has a place in the history of symmetric functions of roots of equations, a man, who for clearness and grasp of material at hand in not only this topic but also in other phases of algebra could well hold his place a century later.
>
> The man was Albert Girard and his work on algebra is a little 34-leaf pamphlet called *Invention Nouvelle en l'Algèbre*, published in 1629.

7.5 A Proof of the Fundamental Theorem

The fundamental theorem of algebra states that every polynomial with real coefficients has a root in the complex numbers. As we discussed in Section 7.2, the theorem's name is misleading. Any proof depends on some information about the real numbers, hence on results from calculus or analysis. At the end of Section 7.2, we mentioned the existence of proofs that are entirely algebraic if we take two results as given: every positive real number has a real square root (Theorem 1.10) and every cubic polynomial with real coefficients has a real root (Theorem 1.14). With the ideas of Section 7.4, we can now describe one such proof, essentially due to Laplace in 1795. That is the purpose of this section.

In addition to Theorems 1.10 and 1.14, our proof of the fundamental theorem will use two results developed in this chapter: Theorem 7.5, which states for a polynomial $f(x)$ with real coefficients that a field K exists containing both $\mathbb{C}$ and a set of roots of $f(x)$, so that $f(x)$ factors in $K[x]$ as a product of linear factors, and Theorem 7.17, the fundamental theorem on symmetric polynomials. We will also need the principle of mathematical induction.

Induction has been used implicitly on occasion, for instance in the proof of de Moivre's formula in Exercise 4.23 and the proof of the two-variable case of the fundamental theorem on symmetric polynomials in Section 7.4. Its application in proving the fundamental theorem is more subtle. Hence, in preparation for that proof, an explicit discussion of induction is warranted.

Let's write $\mathbb{N}$ for the set of positive integers. (This is a traditional mathematical notation, the letter "N" suggesting the natural numbers, another name for the positive integers.) Here is the statement of the principle:

Theorem 7.20 (principle of mathematical induction)**.** *Let S be a subset of $\mathbb{N}$. Assume that S contains* 1 *and that if S contains a positive integer k, then S contains $k + 1$. Then $S = \mathbb{N}$.*

It is perhaps misleading to call the induction principle a theorem. In developing the foundations of arithmetic, we typically adopt the principle as an axiom to be satisfied by the positive integers rather than a theorem to be proved about them. But this is a subject for another book (for instance, [42, pp. 149–150]). Whatever the principle's logical status, we will accept it as a valid statement.

The principle becomes a proof technique when we want to verify a family of statements indexed by the positive integers. De Moivre's formula is an example. We wish to prove for every positive n that

$$(\cos\theta + i\sin\theta)^n = \cos n\theta + i\sin n\theta.$$

Let S be the set of positive integers for which the equality holds. We see immediately that S contains 1. Suppose we can show for any positive integer k that if k is in S, so is $k + 1$. Then the principle of mathematical induction implies that $S = \mathbb{N}$ and de Moivre's formula holds for all positive integers. What we must show, then, is that if

$$(\cos\theta + i\sin\theta)^k = \cos k\theta + i\sin k\theta,$$

then

$$(\cos\theta + i\sin\theta)^{k+1} = \cos(k+1)\theta + i\sin(k+1)\theta.$$

This is essentially what we checked in Exercise 4.23.

Sometimes we wish to prove a family of statements indexed not by $\mathbb{N}$ but by the set $\mathbb{N} \cup \{0\}$ of non-negative integers. The principle of induction works. We prove that the statement associated with 0 is true, then we prove for any non-negative integer k that the validity of the $(k+1)$st statement follows from the validity of the kth statement.

We are ready to prove the fundamental theorem of algebra: any polynomial $f(x)$ of positive degree in $\mathbb{R}[x]$ has a root in $\mathbb{C}$. We might think of proceeding inductively using the degree of $f(x)$ as the induction index. But we already know the fundamental theorem is true for odd-degree polynomials, so induction on degree isn't likely to be a successful strategy. Instead, following Laplace, we do something more clever, applying induction not to the degree of $f(x)$ but to its "degree of evenness."

A positive integer n factors as $2^k s$ for some non-negative integer k and positive odd integer s. What interests us is the exponent k. The smaller k is, the closer n is to being odd. Or, the larger k is, the more even (so to speak) n is. This exponent k is the basis for our induction argument.

The starting point for our induction is the case $k = 0$. The positive integers for which k is 0 are the odd ones. Thus, we begin by considering the odd-degree polynomials in $\mathbb{R}[x]$. By Theorem 1.14, they have real roots, so the fundamental theorem holds for them. Our induction argument is off to a successful start.

We next choose an integer $k \geq 0$ and assume for all odd positive integers s that the fundamental theorem holds for polynomials of degree $2^k s$. This is called the *inductive hypothesis*. Using it, we wish to prove for all odd positive integers t that the fundamental theorem holds for polynomials of degree $2^{k+1}t$. If we do, then by the principle of mathematical induction, we will have shown that the fundamental theorem holds for polynomials of degree $2^\ell u$, for all non-negative integers ℓ and all odd positive integers u. In other words, it holds for all polynomials of positive degree, as desired.

Suppose $f(x)$ is a polynomial in $\mathbb{R}[x]$ of degree n, with n of the form $2^{k+1}t$ and t odd. The induction argument requires us to show that $f(x)$ has a root in $\mathbb{C}$. By Theorem 7.5, there is a field K containing $\mathbb{C}$ that contains roots $r_1, r_2, \ldots, r_n$ of $f(x)$ (repetitions allowed), with $f(x)$ factoring in $K[x]$ as

$$(x - r_1) \cdots (x - r_n).$$

Fix an arbitrary integer m. For a pair of integers i and j satisfying $1 \leq i < j \leq n$, we can form in K the element

$$r_i + r_j + m r_i r_j.$$

Since there are $n(n-1)/2$ pairs i and j, there will be $n(n-1)/2$ such elements. Therefore, the polynomial

$$F(x) = \prod_{i<j}(x - (r_i + r_j + mr_ir_j))$$

in $K[x]$ has degree $n(n-1)/2$.

If we permute the roots $r_1, \ldots, r_n$, we permute the $n(n-1)/2$ elements $r_i + r_j + mr_ir_j$, leaving them unchanged as a set. Therefore we also leave unchanged the coefficients of the polynomial $F(x)$. But these are polynomial expressions in the roots $r_1, \ldots, r_n$ of $f(x)$ with integer coefficients. Therefore, by a suitable application of the fundamental theorem on symmetric polynomials, they are polynomial expressions in the coefficients of $f(x)$ with integer coefficients, which is to say, they are real numbers.

We have proved that $F(x)$ lies in $\mathbb{R}[x]$. Recall that $f(x)$ has degree n, with $n = 2^{k+1}t$ and t odd. Since $k+1$ is positive, n is even. Hence, $n-1$ and $t(n-1)$ are odd. But $F(x)$ has degree $n(n-1)/2$ and

$$\frac{n(n-1)}{2} = \frac{2^{k+1}t(n-1)}{2} = 2^k t(n-1).$$

Thus, the degree of $F(x)$ is exactly of the form for which we assumed by the inductive hypothesis that the fundamental theorem holds: the product of 2^k and an odd number. We conclude that $F(x)$ has a root in $\mathbb{C}$, which means that there is a fixed pair $\{i, j\}$ (depending on m) for which the element $r_i + r_j + mr_ir_j$ of K lies in the smaller field $\mathbb{C}$. At the risk of over-complicating the notation, we might write this element as $r_{i(m)} + r_{j(m)} + mr_{i(m)}r_{j(m)}$ to emphasize its dependence on the choice of the integer m.

We chose m arbitrarily. The argument we just made applies for each choice of m, and there are infinitely many such choices. But there are only $n(n-1)/2$ pairs of integers $\{i, j\}$ between 1 and n. Therefore, there must be distinct integers m and m' yielding the same pair; that is, there exist integers $m \neq m'$ for which $i(m) = i(m')$ and $j(m) = j(m')$.

Let's simplify our notation, writing i and j for the integers in this pair. Then $r_i + r_j + mr_ir_j$ and $r_i + r_j + m'r_ir_j$ are both complex numbers, so their difference $(m - m')r_ir_j$ is as well. Since $m - m'$ is a non-zero integer, we can divide by it to deduce that r_ir_j is complex. Hence mr_ir_j is complex and we can subtract it from the complex number $r_i + r_j + mr_ir_j$ to deduce that $r_i + r_j$ is complex. We have proved that $f(x)$ has two roots r_i and r_j in the field K whose sum $r_i + r_j$ and product r_ir_j lie in the smaller field $\mathbb{C}$.

There is one last step. The polynomial $(x - r_i)(x - r_j)$ equals

$$x^2 - (r_i + r_j) + r_ir_j.$$

Since $r_i + r_j$ and $r_i r_j$ are complex, it lies in $\mathbb{C}[x]$. By design, its roots are r_i and r_j. We proved in Theorem 4.7 that any quadratic polynomial in $\mathbb{C}[x]$ has complex roots. Hence, r_i and r_j lie in $\mathbb{C}$. But r_i and r_j are roots of $f(x)$, so $f(x)$ has roots in $\mathbb{C}$.

We have proved, under the assumption that polynomials whose degree is a product of 2^k and an odd number have roots in $\mathbb{C}$, that polynomials whose degree is a product of 2^{k+1} and an odd number also have roots in $\mathbb{C}$. We also proved the base case, that polynomials of odd degree have roots in $\mathbb{C}$. By the principle of mathematical induction, polynomials of any positive degree have roots in $\mathbb{C}$, proving the fundamental theorem.

Let's review where we used our two assumptions. The existence of real roots for odd-degree polynomials entered the stage at the beginning, as the base case of the induction argument. The existence of real square roots for positive real numbers was used, implicitly, at the end, when we quoted Theorem 4.7 on quadratic polynomials in $\mathbb{C}[x]$ having roots in $\mathbb{C}$. This followed by a completing-the-square argument from the existence in $\mathbb{C}$ of square roots of complex numbers (Theorem 4.6), which followed from the existence of real square roots for positive real numbers.

What a beautiful proof! It depends on several of the ideas we have introduced in this chapter, plus a touch of genius. It is a perfect peak on which to bring our study of polynomials to a close.

References

[1] Niels Abel, Mémoire sur les équations algébriques ou l'on démontre l'impossibilité de la résolution de l'équation générale du cinquième degré, (1824) in *Œuvres Complètes, Volume I*, Grøndahl & Søn, Christiania, 1881, 28-33; translated by P. Pesic in [52, pp. 155–169].

[2] Irving Adler, *The Giant Golden Book of Mathematics*, illustrated by Lowell Hess, Golden Press, New York, 1960.

[3] Lars V. Ahlfors, *Complex Analysis, 3rd ed.*, McGraw-Hill, New York, 1979.

[4] Jean-Robert Argand, *Imaginary Quantities: Their Geometrical Interpretation*, translated by A.S. Hardy with a preface by J. Hoüel, D. Van Nostrand, New York, 1881.

[5] Raymond G. Ayoub, Paolo Ruffini's contributions to the quintic, *Archive for History of Exact Sciences*, 23 (1980) 253–277.

[6] Claude-Gaspard Bachet, *Problèmes Plaisants et Délectables, 4th ed.*, Gauthier-Villars, Paris, 1905.

[7] I.G. Bashmakova, *Diophantus and Diophantine Equations*, translated by Abe Shenitzer with the editorial assistance of Hardy Grant, Mathematical Association of America, Washington, D.C., 1997.

[8] I.G. Bashmakova and G.S. Smirnova, *The Beginnings and Evolution of Algebra*, translated by Abe Shenitzer with the editorial assistance of David A. Cox, Mathematical Association of America, Washington, D.C., 2000.

[9] William P. Berlinghoff and Fernando Q. Gouvêa, *Math Through the Ages*, Oxton House Publishers, Farmington, Maine, 2002.

[10] A.E. Berriman, The Babylonian quadratic equation, *The Mathematical Gazette*, 40 (1956) 185–192.

[11] Rafael Bombelli, *l'Algebra*, Giovanni Rossi, Bologna, 1579.

[12] Girolamo Cardano, *The Great Art; or, The Rules of Algebra*, translated and edited by T. Richard Witmer, The M.I.T. Press, Cambridge, Massachusetts, 1968.

[13] Girolamo Cardano, *The Book of My Life*, translated by Jean Stoner with an introduction by Anthony Grafton, New York Review Books, New York, 2002.

[14] Henry Thomas Colebrooke, *Algebra with Arithmetic and Mensuration from the Sanskrit of Brahmegupta and Bhascara*, John Murray, London, 1817.

[15] S. Cuomo, *Ancient Mathematics*, Routledge, London, 2001.

[16] Abraham de Moivre, The analytic solution of certain equations of the third, fifth, seventh, ninth and other higher uneven powers, by rules similar to those called Cardan's, *Philosophical Transactions*, 25 (1707) 2368–2371; portions translated in [60, pp. 441–444].

[17] Abraham de Moivre, On the reduction of radicals to simpler terms, or the extraction of roots of any binomial $a + \sqrt{+b}$ or $a + \sqrt{-b}$, *Philosophical Transactions*, 40 (1739) 463-478; portions translated in [60, pp. 447–450].

[18] René Descartes, *A Discourse on the Method of Correctly Conducting One's Reason and Seeking Truth in the Sciences*, translated by Ian Maclean, Oxford University Press, Oxford, 2006.

[19] René Descartes, *The Geometry of René Descartes*, translated by David Eugene Smith and Marcia L. Latham, Open Court Publishing Company, La Salle, Illinois, 1952.

[20] Leonard Eugene Dickson, *Elementary Theory of Equations*, John Wiley and Sons, New York, 1914.

[21] Peter Doyle and Curt McMullen, Solving the quintic by iteration, *Acta Mathematica*, 163 (1989) 151–180.

[22] David S. Dummit and Richard M. Foote, *Abstract Algebra, 3rd ed.*, John Wiley & Sons, New York, 2004.

[23] William Dunham, *Euler: The Master of Us All*, Mathematical Association of America, Washington, D.C., 1999.

[24] Gotthold Eisenstein, Allgemeine Auflösung der Gleichung von den ersten vier Graden, *Journal für die reine and angewandte Mathematik* 27 (1844) 81-83; reprinted in *Mathematische Werke, Vol. I*, Chelsea Publishing Company, New York, 1975, 7–9.

[25] Gotthold Eisenstein, Entwicklung von $\alpha^{\alpha^{\alpha}}$, *Journal für die reine and angewandte Mathematik* 28 (1844) 49-52; reprinted in *Mathematische Werke, Vol. I*, Chelsea Publishing Company, New York, 1975, 122–125.

[26] Leonhard Euler, *Elements of Algebra*, translated by Rev. John Hewlett, Springer-Verlag, New York, Berlin, Heidelberg, Tokyo, 1984.

[27] Charles Fefferman, An easy proof of the fundamental theorem of algebra, *American Mathematical Monthly*, 74 (1967) 854–855.

[28] Fibonacci, *Fibonacci's Liber Abaci: A Translation into Modern English of Leonardo Pisano's Book of Calculation*, translated by L.E. Sigler, Springer-Verlag, New York, 2002.

[29] Benjamin Fine and Gerhard Rosenberger, *The Fundamental Theorem of Algebra*, Springer-Verlag, New York, 1997.

[30] H. Gray Funkhouser, A short account of the history of symmetric functions of roots of equations, *The American Mathematical Monthly*, 37 (1930) 357–365.

[31] Évariste Galois, Mémoire sur les conditions de résolubilité des équations par radicaux, in *Écrits et Mémoires Mathémathiques d'Évariste Galois*, Robert Bourgne and J.-P.Azra, eds., Gauthier-Villars & Cie, Paris, 1962, 43–71.

[32] Albert Girard, *Invention Nouvelle en l'Algèbre*, Guillaume Iansson Blaeuw, Amsterdam, 1629.

[33] Anthony Grafton, *Cardano's Cosmos*, Harvard University Press, Cambridge, Massachusetts, 1999.

[34] C. Hermite, Sur la résolution de l'équation du cinquième degré, *Comptes Rend. Acad. Sci. Paris* 46 (1858) 508–515.

[35] Jens Høyrup, *Lengths, Widths, Surfaces*, Springer-Verlag, New York, 2002.

[36] Louis Charles Karpinski, *Robert of Chester's Latin Translation of the Algebra of Al-Khowarizmi, with an Introduction, Critical Notes and an English Version*, MacMillan Company, New York, 1915.

[37] Felix Klein, *Lectures on the Icosahedron and Solutions of Equations of the Fifth Degree*, translated by George Gavin Morrice, Dover Publications, New York, 1956.

[38] Steven G. Krantz, *Real Analysis and Foundations, 2nd ed.*, Chapman and Hall, Boca Raton, Florida, 2005.

[39] J.L. Lagrange, Traité de la résolution des équations numériques de tous les degrés, fourth edition, reprinted from the second edition of 1808, in *Œuvres de Lagrange, Tome Huitième*, Gauthier-Villars, Paris, 1879.

[40] Eli Maor, *Trigonometric Delights*, Princeton University Press, Princeton, New Jersey, 1998.

[41] Barry Mazur, *Imagining Numbers (Particularly the Square Root of Minus Fifteen)*, Farrar, Straus and Giroux, New York, 2003.

[42] Elliott Mendelson, *Introduction to Mathematical Logic, 5th ed.*, Chapman and Hall, Boca Raton, Florida, 2010.

[43] Paul J. Nahin, *An Imaginary Tale: The Story of* $\sqrt{-1}$, Princeton University Press, Princeton, New Jersey, 1998.

[44] Otto Neugebauer, *Mathematische Keilschrift-Texte*, reprinted by Springer-Verlag, New York, Heidelberg, Berlin, 1973. (Originally printed by Julius Springer Verlag, Berlin, 1935.)

[45] O. Neugebauer, *The Exact Sciences in Antiquity*, Princeton University Press, Princeton, New Jersey, 1952.

[46] O. Neugebauer and A. Sachs, eds, *Mathematical Cuneiform Texts*, American Oriental Society and the American Schools of Oriental Research, New Haven, Connecticut, 1945.

[47] Isaac Newton, *Universal Arithmetic, or, a Treatise of Arithmetical Composition and Resolution*, Translated from the Latin by Mr. Ralphson, Revised and Corrected by Mr. Cunn, Illustrated and explained in a series of notes by Rev. Theaker Wilder, W. Johnston, London, 1769.

[48] R.W.D. Nickalls, Viète, Descartes and the cubic equation, *Mathematical Gazette* 90 (2006) 203–208.

[49] Øystein Ore, *Cardano: The Gambling Scholar*, Princeton University Press, Princeton, New Jersey, 1953.

[50] Luca Pacioli, *Summa de arithmetica, geometria, proportioni e proportionalità*, Paganino de Paganini, Venice, 1494.

[51] S.J. Patterson, Eisenstein and the quintic equation, *Historia Mathematica* 17 (1990) 132–140.

[52] Peter Pesic, *Abel's Proof*, The MIT Press, Cambridge, Massachusetts, 2004.

[53] Satya Prakash, *A Critical Study of Brahmagupta and His Works*, The Indian Institute of Astronomical & Sanskrit Research, New Delhi, 1968.

[54] Roshdi Rashed, editor, translator, commentator, *Al-Khwārizmī: The Beginnings of Algebra*, SAQI Books, London, 2009.

[55] Reinhold Remmert, The fundamental theorem of algebra, in *Numbers*, Springer-Verlag, New York, 1990.

[56] Eleanor Robson, *Mathematics in Ancient Iraq*, Princeton University Press, Princeton, New Jersey, 2008.

[57] Paolo Ruffini, *Opere Matematiche di Paolo Ruffini*, 3 volumes, edited by Ettore Bortolotti, Volume 1, Palermo, 1915, Volumes 2 and 3, Edizioni Cremonese, Rome, 1953 and 1954.

[58] J.F. Scott, *The Mathematical Work of John Wallis*, Chelsea Publishing Company, New York, 1981.

[59] Jacques Sesiano, *An Introduction to the History of Algebra: Solving Equations from Mesopotamian Times to the Renaissance*, translated by Anna Pierrehumbert, American Mathematical Society, Providence, Rhode Island, 2009.

[60] David Eugene Smith, *A Source Book in Mathematics*, Dover Publications, Mineola, New York, 1959.

[61] John Stillwell, *Mathematics and Its History, 3rd ed.*, Springer Science+Business Media, New York, 2010.

[62] John Stillwell, Eisenstein's footnote, *The Mathematical Intelligencer* 17 (1995) 58–62.

[63] Niccolò Tartaglia, *Quesiti et Inventioni Diverse*, fascimile reproduction from the 1554 edition, Ateneo di Brescia, Brescia, 1959.

[64] B.L. van der Waerden, *Science Awakening*, translated by Arnold Dresden, P. Noordhoff, Groningen, Holland, 1954.

[65] B.L. van der Waerden, *Geometry and Algebra in Ancient Civilizations*, Springer-Verlag, Berlin, New York, 1983.

[66] B.L. van der Waerden, *A History of Algebra from Al-Khwārizmī to Emmy Noether*, Springer-Verlag, Berlin, New York, 1985.

[67] V.S. Varadarajan, *Algebra in Ancient and Modern Times*, American Mathematical Society, Providence, Rhode Island, 1998.

[68] François Viète, *The Analytic Art*, translated by T. Richard Witmer, The Kent State University Press, Kent, Ohio, 1983.

[69] John Wallis, *A Treatise of Algebra, both Historical and Practical*, John Playford, London, 1685.

[70] Caspar Wessel, On the analytical representation of direction, *Memoirs of the Royal Danish Academy of Sciences* 1799; portions translated in [60, pp. 55–66].

Index

About the Author

Ron Irving is a mathematics professor at the University of Washington in Seattle. He was born in suburban New York City, studied mathematics and philosophy at Harvard, and received his Ph.D. in mathematics at MIT. Following a postdoctoral position at Brandeis and a National Science Foundation postdoctoral fellowship year at the University of Chicago and UC San Diego, Irving came to Seattle. He has been a visiting faculty member at UCSD and Aarhus and a member of the Institute for Advanced Study in Princeton. His research interests have ranged over several areas of algebra, including ring theory and the representation theory of Lie groups and Lie algebras.

When Irving began teaching the department's senior algebra course for majors planning on secondary teaching careers, he developed an interest in the preparation of pre-service and in-service teachers. His work with this audience led to receipt of the university's Distinguished Teaching Award and to his book *Integers, Polynomials, and Rings*.

Irving spent seven years in academic administration, serving as department chair for a year, divisional dean of natural sciences for four, and interim dean of the College of Arts and Sciences for two. During this time, he established the Summer Institute for Mathematics at UW, a six-week residential program that brings talented high school students in the Pacific Northwest to the university to share in the excitement of doing mathematics. He continues to serve as the program's executive director. He also began the planning for a new undergraduate degree in integrated sciences designed to meet the needs of future secondary science teachers.

Since returning to the department, Irving has taught a variety of algebra courses and written this book. He has been the director of the Integrated Sciences program, completing its planning, approval, and implementation. He serves as the secretary-treasurer of the Astrophysical Research Consortium, which oversees Apache Point Observatory in the Sacramento Mountains of New Mexico, and is president of the board of the Burke Museum of Natural History and Culture.

Irving is a member of the Mathematical Association of America and the American Mathematical Society. As this book goes to press, he is beginning a second stint as department chair.